Anonymous

Florida

It's Climate, Soil, Productions, And Agricultural Capabilities

Anonymous

Florida
It's Climate, Soil, Productions, And Agricultural Capabilities

ISBN/EAN: 9783744744447

Printed in Europe, USA, Canada, Australia, Japan

Cover: Foto ©berggeist007 / pixelio.de

More available books at **www.hansebooks.com**

ITS

CLIMATE, SOIL, PRODUCTIONS,

AND

AGRICULTURAL CAPABILITIES.

WASHINGTON:
GOVERNMENT PRINTING OFFICE.
1882.

FLORIDA: ITS CLIMATE, SOIL, PRODUCTIONS, AND AGRICULTURAL CAPABILITIES.

Florida, from its first discovery in 1512, has been in an unsettled condition, conquered and reconquered, ceded and receded, harassed by Indian wars, and when just entering on a period of stability and prosperity plunged into a civil contest, which decimated and impoverished her people. Ceded to the United States in 1821, and admitted as a State in 1845, her resources have remained latent and undeveloped, and her 60,000 square miles of territory comparatively wild and uninhabited until about the close of the late civil war. Since that period the intelligence of the world has been directed to this favored land, and thousands have annually sought health and pleasure and new homes within her borders. Other thousands will come, when informed of the advantages and attractions of this productive semi-tropical State, only awaiting capital and industry to render it one of the wealthiest and most prosperous of the Federal Union.

The peninsular portion of the State, known as East and South Florida, is some 300 miles in length from north to south, and averages about 100 in width, gradually narrowing towards its southern terminus. The Gulf Stream on its eastern coast causes the trade winds of the Atlantic to sweep over the land from east to west by day, while the returning cool breezes from the Gulf refresh the land by night. These daily breezes constantly purify and vivify the atmosphere, and prevent oppressive heat or sultriness.

Generally the lands bordering on the ocean and Gulf are level and at no great elevation above tide-water; midway there is a table-land elevation, reaching nearly to the everglades. The extreme southern portion is low, though, from recent surveys, it is found that it can be effectually drained and made available for cultivation.

No State in the Union has such an extent of coast, being nearly 1,200 miles in length, extending from Fernandina on the north to Pensacola on the west, indented every few miles by large bays, running inland in many places from ten to thirty miles, with large rivers like the Saint John's, Oclawaha, Kissimmee, Indian, Halifax, Saint Mary's, Suwannee, and Apalachicola, navigable from north to south, and easterly and westerly between the Gulf and Atlantic Ocean. There are other connecting navigable streams in all parts of the State, and lakes, large and small, scattered and grouped together, all of which furnish local transportation facilities, and abound in excellent varieties of fish; many connect with navigable streams, and all can be easily connected by short canals or railroads with each other and the great arteries of water leading to the sea and Gulf. The interior lakes of Florida, large and small, is one of its remarkable features. The largest of these is Lake Okee-chobee, situated in the middle of the table-lands of the peninsula, and contains an area of 1,000 square miles, without visible outlet to the sea.

The soil in the greater portion of the State is sandy, except in the hill lands and hammocks, where large portions of clay and alluvium are found. The sand is not the sharp, siliceous sand of the ocean, or resembling the sandy lands of other States; this soil has more or less of loam and a large percentage of lime and organic remains, giving it much fertility. The country is well watered, not only by its larger and smaller rivers and lakes, but by innumerable creeks and springs. Springs of great volume are found in every portion of the State, some of such magnitude that they form navigable rivers from their source; of such are the Blue Springs, in Jackson County, in the west; Wakulla Springs, in Wakulla County; Blue Springs, in Hernando County, in the middle; Silver Springs, in Marion County, in the east; the very large Blue Spring on the Saint John's, in Volusia County; the Green Cove Springs, in Clay County, on the shore of the Saint John's; also Clay Spring, in Orange County. Some of these are medicinal—white sulphur, iron, &c. Good water, so universally desired, is found easily at a depth of from eight to fifty feet, according to locality, generally from twelve to twenty feet, but, through the country, the many lakes, and springs, and branches afford ample supply for house and farm purposes. If cistern water is preferred, the average rainfall, being from forty-eight to fifty-four inches annually, insures a supply. The distribution of rivers, creeks, lakes, and springs is not only large, but remarkably uniform throughout the State. Pine lands (pitch and yellow pine) form the basis of Florida. These lands are usually divided into three classes, denoting first, second, and third rate pine lands.

That which is denominated "first-rate pine land" in Florida has nothing similar to it in any of the other States. Its surface is covered for several inches deep with a dark vegetable mold, beneath which, to the depth of several feet, is a chocolate-colored sandy loam, mixed for the most part with limestone pebbles, and resting upon a substratum of marl, clay, or limestone rock. The fertility and durability of this description of land may be estimated from the well-known fact that it has in several districts yielded during fourteen years of successive cultivation, without the aid of manure, four hundred pounds of Sea Island cotton to the acre. These lands are still as productive as ever; so that the limit of their durability is yet unknown. The "second-rate pine" lands, which form the largest proportion of Florida, are all productive. These lands afford fine natural pasturage; they are heavily timbered with the best species of pitch and yellow pine. They are for the most part high, rolling, healthy, and well watered. They are generally based upon marl, clay, or limestone. They will produce for several years without the aid of manure, and, when fertilized, they will yield two thousand pounds of the best quality of sugar to the acre, or about three hundred pounds of Sea Island cotton. They will also produce rice, tobacco, oats, corn, rye, and all the vegetable varieties, as well as the various tropical fruits, which render them more valuable than the best bottom lands in the more northern States.

The lands of the "third rate," or most inferior class, are by no means worthless under the climate of Florida. This class of lands may be divided into two orders; the one comprising high, rolling, sandy districts, which are sparsely covered with a stunted growth of "blackjack" and pine; the other embracing low, flat, swampy regions, which are frequently studded with "bay galls," and are occasionally inundated, but which are covered with luxuriant vegetation and generally with valuable timber. The former of these, it is now ascertained, owing to its calcareous soil, is well adapted to the growth of the Sisal hemp, which is a valuable tropical production. This plant (the *Agava Siciliana*), and the *Agave Mexicana* hemp, also known as the maguay, the pulque plant, the century plant, &c., have both been introduced into Florida, and they both grow in great perfection on the poorest lands of the country. As these plants derive their chief support from the atmosphere, they will, like the common air-plant, preserve their vitality for many months when left out of the ground.

These lands, besides being valuable for their timber and the naval stores which they will produce, afford an excellent range for cattle, and are susceptible of cultivation in the various productions, when properly ditched and drained.

There is one general feature in the topography of Florida which no other country in the United States possesses, and which affords a great security to the health of its inhabitants. It is this: that the pine lands which form the basis of the country, and which are almost universally healthy, are nearly everywhere studded, at intervals of a few miles, with hammock lands of the richest quality. These hammocks are not, as is generally supposed, low, wet lands; they never require ditching or draining; they vary in extent from 20 acres to 40,000 acres, and will probably average about 500 acres each. Hence, the inhabitants have it everywhere in their power to select residences in the pine lands, at such convenient distances from the hammocks as will enable them to cultivate the latter without endangering their health, if it should so happen that any of the hammocks proved to be less healthy than the pine woods.

Experience in Florida has satisfactorily shown that residences only half a mile distant from cultivated hammocks are entirely exempt from malarial diseases, and that the negroes who cultivate those hammocks and retire at night to pine-land residences, maintain perfect health. Indeed, it is found that residences in the hammocks themselves are generally perfectly healthy after they have been a few years cleared. Newly-cleared lands are sometimes attended with the development of more or less malaria. In Florida the diseases which result from these clearings are generally of the mildest type (simple intermittent fever), while in nearly all the Southern States they are most frequently of a severe grade of bilious fever.

The topographical feature here noted, namely, a general interspersion of rich hammocks, surrounded by high, dry, rolling, healthy pine woods, is an advantage which no other State in the Union enjoys; and Florida forms in this respect a striking contrast with Louisiana, Mississippi, and Texas, whose sugar and cotton lands are generally surrounded by vast alluvial regions, subject to frequent inundations, so that it is impossible to obtain within many miles of them a healthy residence.

The lands which in Florida are, *par excellence*, denominated "rich lands," are, first, the " swamps lands"; second, the "low hammock lands"; third, the "high hammocks"; and fourth, the "first-rate pine, oak, and hickory lands."

The swamp lands are unquestionably the most durably rich lands in the country. They are the most recently formed lands, and are still annually receiving additions to their surface. They are intrinsically the most valuable lands in Florida, being as fertile as the hammocks and more durable. They are evidently alluvial and of recent formation, and occupy natural depressions of basins which have been gradually filled up by deposits of vegetable débris, &c., washed in from the adjacent and higher lands. Ditching is indispensable to all of them in their preparation for successful cultivation. Properly prepared, however, their inexhaustible fertility sustains a succession of the most exhausting crops with astonishing vigor. The greatest yield of sugar ever realized in Florida was four hogsheads per acre, produced on a plantation on Indian River, on which is now located the large orange grove which has given character to the oranges of Florida—the Dummitt Grove, recently purchased by an Italian nobleman, the Duke de Castelluccia. Sugar cane is here instanced as a measure of the fertility of the soil, because it is one of the most exhausting crops known, and is generally grown without rest or rotation. It is not, however, a fair criterion by which to judge of the relative fertility of lands situated in different climates, for we find on the richest lands in Louisiana the crop of sugar per acre is not more than one hogshead, or about half that of East Florida.

This great disparity in the product of those countries is accounted for, not by any inferiority in the lands of Louisiana or Texas, but from the fact that the early incursions of frost in both these States render it necessary to cut the cane in October, which is long before it has reached maturity, while in East Florida it is permitted to stand without fear of frost till December, or till such time as it is fully matured. It is well known that it "tassels" in East Florida, and it never does so in either Louisiana or Texas. When cane "tassels" it is evidence of its having reached full maturity.

In consequence of the heavy outlay of capital required in the preparation of this description of land for cultivation, and from the facility of obtaining hammock land, which requires no ditching or draining, swamp lands have been but little sought after by persons engaged in planting in Florida, and there is now at least a million of acres of the best description of this land vacant in the country, which can be secured at less than two dollars per acre.

Low hammocks, from the fact of their partaking of the nature of hammocks and swamps, and sometimes termed "swammock," are not inferior to swamp lands in fertility, but perhaps are not quite as durable. They are nearly always level, and have a soil of greater tenacity than that of the high hammocks. Some ditching is necessary in many of . them. The soil is always deep. They are also extremely well adapted to the growth of the cane, as has been well attested by the many plantations which were formerly in operation here on this description of land. There is not so large a proportion of low hammock as there is of swamp lands.

High hammocks are the lands in the greatest repute in Florida. These differ from low hammocks, in occupying higher ground, and in generally presenting an undulating surface. They are formed of a fine vegetable mold, mixed with a sandy loam, in many places two feet deep, and resting in most cases on a substratum of clay, marl, or limestone. It will be readily understood by any one at all acquainted with agriculture, that such a soil, in such a climate as Florida, must be extremely productive. This soil scarcely ever suffers from too much wet, nor does drought affect it in the same degree as other lands. High hammock lands produce, with but little labor or cultivation, all the crops of the country in abundance. Such lands have no tendency to break up in heavy masses, nor are they infested with pernicious weeds or grasses. Their extraordinary fertility and productiveness may be estimated by the fact that in several well-known instances in Marion County three hogsheads of sugar have been made per acre on this description of land, after it had been in cultivation six years in successive crops of corn, without the aid of manure.

To sum up its advantages, it requires no other preparation than clearing and plowing to fit it at once for the greatest possible production of any kind of crop adapted to the climate. In unfavorable seasons it is much more certain to produce a good crop than other kinds of land, from the fact that it is less affected by exclusively dry or wet weather. It can be cultivated with much less labor than any other lands, being remarkably mellow, and its vicinity is generally high and healthy. These reasons are sufficient to entitle it to the estimation in which it is held over all other lands. There are, besides the lands already noticed, extensive tracks of savanna lands, which approximate in character, texture of the soil, and period and mode of formation, to the swamp lands, differing only in being destitute of timber. These savannas yield an immense crop of grass, which if cut and properly cured would afford a quality of hay equal to the marsh hay of the Northwest.

In Middle Florida, the counties of Leon, Gadsden, Jefferson, and Madison have large quantites of high, rolling hammock lands, also the county of Jackson in West Florida. They are more undulating than those in East Florida, and are underlaid with a stiff red clay. They are by far the best lands in the State for short-staple cotton, to which they have been almost exclusively appropriated, and to wheat, rye, oats, corn, tobacco, &c. In Volusia County there is a range of low hammock a little back from the coast, from a half to two miles wide, and extending from the head of the Halifax to the head of the Indian River, some 50 miles, as well adapted to sugar culture as any land in the State. The Gulf Hammock, in Levy County, comprises perhaps the largest body of rich land in Florida. It was bought up years ago at from $5 to $10 per acre, by private parties, by whom it is mostly held at the present time. The Atlantic, Gulf and West India Transit Railroad runs through it, and it will no doubt become at an early day one of the garden spots of the State. The clearing of the hammocks, however, is expensive, and, as in every new country, we may expect to see the lands more lightly timbered first brought into cultivation.

CLIMATE.

The climate of Florida, from parallel 31, its northern boundary, to 29, corresponds with that of Portugal, south of Oporto; the southern section of Spain; Oran, Algiers, and Tunis, on the northern coast of Africa; the southernmost part of Italy; the islands of Sicily, Greece, Morea; the isles of the Archipelago, and those of Candia, Rhodes, Cyprus, &c.; Asia Minor, Syria, Mesopotamia, and Armenia. From latitude 29 to 25, bordering almost on the tropics, and including the remainder of East Florida, containing about 33,000,000 acres of land, there is no parallel climate in Europe or Asia Minor.

As climate, in its effects upon the health and vigor of mankind, is of fundamental importance, and enters more largely into the considerations connected with the industrial occupation and development of the country, and as Florida is receiving more special attention from the migratory world than common, I shall be pardoned if I occupy some space in giving the results of scientific investigations in regard to the constituent elements of the climate of this semi-tropical region.

Dr. C. J. Kenworthy, of Jacksonville, a gentleman of extensive experience, as well as practical research, in an address before the Medical Association of Florida, in 1880, presents a more thorough analysis of the constituent elements of the atmosphere and climate of Florida than has yet been given to the public, and I avail myself of his work, and embody the results in liberal extracts.* He says:

The climate of Florida is not excelled by that of any of the United States, and it may be doubted whether it can be equalled elsewhere in the world. Located on the very borders of the torrid zone, and, therefore, so far as latitude alone is concerned, entitled to rank among the hottest portions of the western continent, still her situation between the Gulf of Mexico and the Atlantic is such that, owing to her peculiar form, she is swept alternately by the winds of eastern and western seas, and relieved from those burning heats with which she would otherwise be scorched: and thus it happens that by the joint influences of latitude and peculiar location, she is relieved on the one hand from the rigors of the winter climate of the Northern and Middle States, and on the other from the extreme heat by which not only the other Southern States, but in the summer time the Northern States, are characterized.

While in winter the Northern and Middle States are covered with snow, and frost penetrates the earth to the depth of several feet, and the leafless trees wave their bare and skeleton fingers in the wintry wind, in Florida most of the trees and shrubs are in full foliage, and the winter gardens are filled with vegetables in their most thrifty growth.

In the Northern States the frosts of November and December most effectually put a stop to all agricultural operations, and the farmer is compelled to feed his stock for from four to six months, and is himself confined to the getting of fuel and lumber, thus in one portion of the year consuming a large share of the result of his labor in the other

But in Florida this very winter season is better adapted to building, clearing land, and the performance of all necessary extra work on the farm than even the summer.

In the North all regular farming work is of necessity crowded into the space of less than half the year, while in Florida there is scarcely a single day in the whole year that may not be devoted to purely agricultural work.

In some of the Northern States the mean average range of the thermometer is from 30° below zero to 90° and 100° above. In Florida, for many years, the range of the thermometer has been less than half as great.

The word "climate" in its common signification indicates a region bounded by certain arbitrary lines, but in medicine it possesses a wider meaning. The effect of climate upon the human system is the sum of the influences which are connected with many factors. The climate of any locality, professionally speaking, depends upon its temperature, atmospheric vicissitudes, prevailing winds, humidity, its elevation above the sea level, its proximity to the ocean or oceanic currents, its contiguity to mountains, lakes, rivers, arid areas, soil, drainage, vegetable productions, malaria, general sanitation and other factors, which we shall briefly consider.

Considering climatic factors, as a result of experience, observation, investigation, and study, we are convinced that Florida presents more attractions and advantages as a winter resort for invalids than any State in the Union. The temperature is favorable, the mean relative humidity is peculiarly adapted to the treatment of all forms of pulmonary disease, the air is salubrious, and in a large portion of the State dry and bracing; atmospheric changes are infrequent, and not so great as in other sections east of the Rocky Mountains. Rains are infrequent, and sunshine and fine weather the rule. The State possesses insular, interior, dry and moist localities, semi-tropical and cooler sections; and if the nature of any given case should necessitate a change of base, a suitable climate can be reached in a few hours and at a trifling expense.

Dr. Charles A. Lee, the learned editor of Copeland's Medical Dictionary, remarks:

Proceeding south from Canada to Florida, the seasons become more uniform in proportion as their annual temperature increases, and they glide imperceptibly into each

---

*Climatology of Florida, by C. J. Kenworthy, M. D., of Jacksonville. Reprint from Transactions Florida Medical Association, session 1880.

other, exhibiting no great extremes. Compared with the other regions of the United States, the peninsula of Florida has a climate wholly peculiar. The climate is so *exceedingly mild and uniform* that, besides the vegetable productions of the Northern States generally, many of a tropical character are produced. We have already spoken of the mildness of the climate of this region; it appears to possess an insular temperature not less equable and salubrious in winter than that afforded by the South of Europe, and is, therefore, well adapted to those forms of pulmonary disease—as bronchitis and incipient phthisis—as are benefited by a mild climate. *Mildness and uniformity* are the two distinguishing characteristics of the Florida peninsula. If we compare the climate of East Florida with the most favored situations on the continent of Europe, and the islands held in the highest estimation for mildness and equability of temperature, in regard to the mean temperature of winter and summer, that of the warmest and coldest months, and that of successive seasons, we shall find the results generally in favor of the former.

After citing the mean difference of successive months and annual range of a number of climatic. resorts in comparison with stations in Florida, he remarks:

Thus it is easily demonstrated that invalids requiring a mild winter residence have gone to foreign lands in search of what might be found at home—an evergreen land, in which wild flowers never cease to unfold their petals.

In discussing the most suitable climates for invalids, Dr. Wilson, late medical inspector of camps and hospitals, United States Army, remarks:

Neither upon the southern coast of France, nor anywhere under the bright Italian skies, can a winter climate be found so equable and so genial to the delicate nerves of most invalids as can be enjoyed in our sanitary stations in Florida.

Temperature is an important factor in climate, and Dr. A. S. Baldwin, of Jacksonville, in an address before the medical association, gives tabular statements of mean temperature at eighteen different stations in Florida during a term of years, including his own observations, which extend through a period of thirty-six years.

From this table it appears that the mean temperature for the entire State is: For spring, 71° 62'; summer, 80° 51'; autumn, 71° 66'; winter, 70° 05'.

For stations on and south of latitude 28° north: Spring, 74° 94'; summer, 81° 93'; autumn, 76° 27'; winter, 63° 69'; and for the year, 74° 87'.

For the stations north of latitude 28° north: Spring, 70° 66'; summer, 80° 10'; autumn, 70° 23'; winter, 58° 29'; and for the year, 69° 82'.

During the spring the temperature south of latitude 28° north is 4°. 28'; for summer, 1° 83'; for autumn, 6° 34'; and for winter, 5° 40' higher than it is north of 28°; showing that the difference between the summer and winter temperature is less south than north of parallel 28. After explaining the astronomical law which governs, the doctor remarks:

Although there is more absolute heat at Jacksonville, Fla., latitude 30°, during the entire year than there is at Milwaukee, Wis., latitude 43° 03', yet there is more heat received from the sun at the latter place during the three summer months than at the former during the same period, and Wisconsin is liable to experience a higher temperature during the summer months than is Florida during the same time.

The comparative humidity of Florida, as connected with health, is shown in the appended tables, presented in the valuable address of Dr. C. J. Kenworthy, of Jacksonville, before cited :

The mean relative humidity of the localities referred to, for the cold months, is as follows :

Localities.	November.	December.	January.	February.	March.	Mean for 5 months.	Mean for 5 months.
	Per cent.	Per cent.	Per cent.	Per cent.	Per cent.	Per cent.	Per cent.
*Cannes and Mentone	71. 8	74. 2	72. 0	70. 7	73. 3	72. 4	
Augusta, Ga	71. 8	72. 6	73. 0	64. 7	62. 8	68. 9	
Breckinridge, Minn	76. 9	83. 2	76. 8	81. 8	79. 5	79. 6	}
Duluth, Minn	74. 0	72. 1	72. 7	73. 3	71. 0	72. 6	} 74. 3
Saint Paul, Minn..............,	70. 3	73. 5	75. 2	70. 7	67. 1	71. 3	}
Jacksonville, Fla.............,	71. 9	69. 3	70. 2	08. 5	63. 9	68. 8	}
Key West, Fla ....,	77. 1	78. 7	78. 9	77. 2	72. 2	76. 8	} 72. 7
Punta Rassa, Fla.............	72. 7	73. 2	74. 2	78. 7	69. 9	72. 7	}

From the above data it appears that the mean relative humidity of Cannes and Mentone, during the cold months, exceeds that of Jacksonville by nearly 4 per cent. Three stations in Minnesota have a mean of 74.3, and three in Florida a mean of 72.7, showing a per cent. of 1.6 in favor of Florida, and 5.5 per cent. in favor of Jacksonville over Minnesota.

If we take the entire year, for a period of five years, we will find but little difference in the mean relative humidity of Minnesota and Florida, as the following data, kindly furnished us by the Chief Signal Officer of the United States Army, will demonstrate :

Years.	Minnesota.			Florida.		
	Breckinridge.	Duluth.	Saint Paul.	Jacksonville.	Punta Rassa.	Key West.
	Per cent.	Per cent.	Per cent.	Per cent.	Per cent.	Per cent.
1875......................................	75. 7	67. 2	69. 0	70. 3	76. 0	71. 5
1876......................................	07. 7	68. 2	69. 1	67. 2	73. 9	76. 1
1877......................................	72. 2	71. 9	67. 6	69. 3	70. 5	74. 1
1878......................................	76. 2	71. 5	67. 7	68. 7	72. 4	74. 5
1879......................................	74. 1	72. 8	05. 8	60. 7	72. 3	74. 2
Mean for five years..............	73. 2	70. 3	67. 7	69. 0	73. 0	74. 2
Mean for five years for States. ..	....	70. 4		....	72. 1	....

Thus it will be perceived that Jacksonville possesses a lower mean relative humidity than most of the celebrated winter resorts. There is one important point to which I wish to direct your attention, and that is the remarkably low percentage of humidity during the much dreaded month of March—the mean for five years being but 63.9, as against 73.3 at Mentone, 76.8 at Atlantic City, 79.5 at Breckinridge, Minn , 68.4 at Nassau, N. P., and 67.1 at Saint Paul. When referring to the effects of change of climate, Dr. Madden remarks : "The temperature of a locality, to which so much importance is properly assigned, is no criterion of its climate as a health resort, the fact being that invalids are painfully sensible of variations in the *hygrometric state of the atmosphere*, which are in no way indicated by the thermometer, so much relied on."

The thermometric range in this city is not too high nor too low. As evidence of this, we find the mean temperature for the *five* cold months, for five years, to be for November, 62°.1; December, 55°.8; January, 56°.2 ; February, 56°.9 ; March, 62°.7 ; mean for five months, 58°.7.

Dr. Lente, p. 17, when discussing temperature as a constituent of climate, and referring to certain winter resorts north of Florida, remarks : "A mean winter temperature of about 48° is too low to *entice* many of the feeble invalids out of doors, except in calm and sunshiny weather. But in some of them this degree of cold is enhanced, as far as the sensations are concerned, by the winds which frequently prevail. At such times, most invalids will, therefore, be found hovering over the comfortable wood fires, and will be pretty sure to keep all the apertures of their chambers closed at night, thus depriving themselves, during by far the greater part of the twenty-four hours, of the principal means of cure. In Florida the sun shines so brightly, the air is so balmy, the songs of the birds so enlivening, and the orange trees in their bloom, or laden with their golden fruit, lend such a charm to the outlook from the windows, that the most indolent or the most cold-blooded invalid feels little inclined to stay in doors."

Prof. J. Hughes Bennett, of the University of Edinburgh, remarks : "And when, in our own country, the question arises, Where shall we send the consumptive patient in order to avoid our changeable climate and cold winter winds in winter? we naturally say : To a land where, during that portion of the year, the weather is warm and equable. Much has been written on climate, but the one which appears to me best is that which will enable the phthisical patient to pass a few hours every day in the open air, without exposure to cold or the vicissitudes of temperature on the one hand, or excessive heat on the other."[*]

With the exception of the month of December, fogs are seldom seen, and when they do occur they are not dense, and disappear as soon as the sun appears above the horizon. Interested parties have intimated that the city of Jacksonville is a "rainy locality," and, in consequence, an "unsuitable place for invalids." To correct this error, we will furnish data applicable to a few winter resorts, from reliable sources :

RAINFALL IN INCHES AND HUNDREDTHS.

Localities.	Number of years.	November.	December.	January.	February.	March.	Five months.
Nice	28	5.11	4.12	3.06	1.68	2.89	16.86
Mentone	9	5.34	3.15	1.70	2.18	4.13	16.50
Nervi	7	6.00	4.88	4.78	2.33	4.49	23.40
Genoa	29	7.61	4.86	4.39	4.27	3.59	24.72
Atlantic City, N. J	5	4.61	3.60	2.76	2.10	3.86	16.93
Augusta, Ga	5	4.56	3.09	3.70	3.64	5.65	20.64
Jacksonville, Fla	5	3.02	3.38	2.34	5.14	2.84	16.62
Key West, Fla	5	2.43	1.33	2.18	2.22	0.94	9.10
Punta Rassa, Fla	5	2.38	0.99	1.69	2.67	1.04	8.77

We have referred to sunshine as an important aid in the treatment of disease and broken health, and as an evidence that Florida is favored with sunshine and fair weather, and not "cloudy, foggy, and rainy weather," as interested parties have as-

* Bennett's Practice of Medicine, pp. 326, 476.

serted, I shall direct your attention to the following meteorological data, furnished by
J. W. Smith, the observer in charge of this station:

METEOROLOGICAL DATA FROM SIGNAL OFFICE, U. S. A., JACKSONVILLE, FLA.

Date.	Rainy days.					Remarks.
	November.	December.	January.	February.	March.	
1874–1875	14	6	15	10	5	⎫
1875–1876	10	4	4	8	7	⎪ "Rainy days," all days on
1876–1877	5	10	6	6	6	⎬ which rain fell.
1877–1878	9	9	5	10	8	⎪
1878–1879	5	8	5	9	3	⎭
Average	8.6	7.4	7	8.6	5.8	37.4 days in five months.

Date.	Cloudy days.					
	November.	December.	January.	February.	March.	
1874–1875	4	6	12	3	8	
1875–1876	8	4	5	7	2	
1876–1877	6	8	2	10	7	
1877–1878	10	11	11	9	10	
1878–1879	9	11	5	11	4	
Average	7.4	7	7	8	6.2	35.6 cloudy days five months.

J. W. SMITH,
Observer in Charge.

When a day is marked "rainy," but a few drops may have fallen, and it is no evidence that the entire day was rainy. A measurable or a non-measurable quantity
may fall in a few minutes, and the remainder of the day be bright and clear. In
Florida the rains are frequently "torrential, in short, severe bursts," followed by
bright and clear weather. For the purposes of comparison, we will refer to the number of rainy days during the cold months in Jacksonville, Mentone, and Saint Paul.

RAINY DAYS (INCLUDING SNOW).

Locality.	Years.	November.	December.	January.	February.	March.	Five months.
Jacksonville	5	8.6	7.4	7	8.6	5.8	37.4
Mentone	8	10.1	7.25	5.1	5.66	9.55	37.48
Saint Paul	1	4	13	8	6	11	42

Ozone is considered an important constituent of the atmosphere, for
by its presence pure air may be inferred to exist. This allotropic condition of oxygen possesses great power of destruction of organic matter
floating in the atmosphere. The Florida peninsula is surrounded by
the sea, and the land is almost constantly fanned by sea-breezes, containing a large amount of ozone. According to the researches of Burdel,

he found as "much ozone in the air of marshes as in other air." Clemens
says: "There is a large proportion of oxygen near the surface of lakes,
giving the reaction of ozone," more especially if there are certain aquatic ·
plants present; and he also remarks that at some feet above the surface
the reaction is lost. Grallois has lately stated that he "found more
ozone over marshes than anywhere else." Dr. Schreiber, of Vienna,
asserts "that the turpentine exhaled from pine forests possesses, to a
greater degree than all other substances, the property of converting
the oxygen of the air into ozone." In this connection, Dr. Denison
remarks: "If this be true, it will explain why a residence among the
balsamic odors of the pines has long been esteemed of benefit to the
pulmonary invalid." Florida is densely covered with pine forests, and
if we accept the statement of Dr. Schreiber, Dr. Jones is in error. Dr.
Moffat found the quantity of ozone in the atmosphere greater when the
mean daily temperature was above the mean. According to the researches of Dr. Denison in Colorado, the excess of ozone appeared during the spring months on the plains, and came proportionately later in
the season the higher up the observations were made. Says Dr. Kenworthy:

Malaria is a subject which enters into the discussion of all southern climes, and we
unhesitatingly assert that Florida has been misrepresented in this respect. "It is the
custom," remarks Dr. Lente, p. 21, "of many persons living at Florida resorts, off the
Saint John's River, to represent, for obvious reasons, that fever prevails there the year
round, and that it is dangerous to resort to it at any time. In this manner they have
excited senseless alarm in the minds of those proposing to come to Florida, and have
diverted them to other Southern resorts, thus in the end injuring themselves as well
as others." Unprincipled hotel keepers and runners, and the agents of steamboat and
railroad lines leading to other localities, aid more or less in this fraudulent attempt to
gain patronage. The bugbear, malaria, is, in my humble opinion, a prolific source of
disease among visitors to Florida. By misrepresentations (to use a mild term) tourists
and invalids have been led to believe that the entire water supply is productive of
disease, and as a consequence they refrain from drinking a sufficient quantity of water,
or dilute it with poor whisky or brandy to counteract its bad effects Interested parties have expatiated so much with regard to the air being charged with malaria in
winter, that invalids and patients become alarmed, and as a sequence they daily swallow quinine, and thereby produce nervous or functional derangements. They keep
the pure air out of their rooms, breathe an air contaminated with their own breaths
and exhalations, and at night assemble in halls and parlors and inhale vitiated air
poisoned by their own breaths and the elements resulting from the combustion of coal-
gas and kerosene. They inhale for hours at a time air charged with carbonic acid,
and shun the pure night air as they would the emanations of the deadly Upas tree.
Visitors act imprudently, and as a consequence suffer from nervous derangements,
colds, and diarrhœas, which they attribute to malaria or the climate. The cause of
slight indispositions affecting visitors is not malaria, but indulgence at table, change
of drinking water, eating excessive quantities of fruit, or the inhalation of air poisoned
by human breaths, or the resultants of the combustion of coal-gas and kerosene, and a
deficiency of the pure air that a beneficent Creator has placed everywhere within their
reach. If visitors would let quinine and arsenical pills alone, control their appetites,
eat moderately, inhale plenty of the salubrious air of the State, and not swelter in
heated halls, parlors, and unventilated bed-rooms, we should hear less of the bugbear,
malaria.

At various times since 1844, I have navigated the larger streams of this State, visited the everglades and Lake Okeechobee, and almost every bay, inlet, and river, from Cape Sable to the Suwannee River, and for over two months at a time slept in an open boat, with nothing but a simple awning stretched over the boat's boom, and in no instance did my companions or self suffer from malaria or a chill. Before I became a resident of the State, my companions and self were unacclimated, and in no instance were we so foolish as to swallow quinine, arsenic, or alcoholic liquors as antidotes to malaria or chills. I speak from personal observation, experience, and extended inquiry in various portions of the State, and I unhesitatingly assert that the opinion entertained with regard to the prevalence of malaria during the cold months in Florida is unfounded. When discussing the advantages of Florida as a climatic resort, the eminent Dr. Forry predicted, "from a long residence in Florida, attached to the United States Army, that when the period of the red man's departure shall have passed, the climate of this land of flowers would acquire a celebrity as a winter residence not inferior to that of Italy, Madeira, or Southern France." *

"All know," remarks Dr. Brinton, p. 128, "how terribly arduous must be campaigning among the everglades of Florida, yet the yearly mortality from disease of the Regular Army there was only 26 per 1,000 men; the average of the Army elsewhere was 35 per 1,000, while in Texas it rose to 40, and on the Lower Mississippi to 45 per 1,0 0." If persons are suffering from malarial cachexia they may have chills at any season in any climate. But a few weeks since I was requested to visit a young lady visitor, whose home is Fifth Avenue, New York. The only time she had been dressed for three months was the day she was driven to the Savannah steamer. Upon inquiry I found that quinine, arsenic, and Warbeck's tincture had failed to cure her of chills. She arrived in this city in the latter part of February, and at the end of two weeks she departed for home, sans chills, sans malaria, sans debility. From my experience in hospitals and private practice in and near New York, I have no hesitation in stating that malarious diseases are more frequent there than in Florida. From my observations from Canada to the Gulf of Mexico, I am convinced that febrile diseases assume a milder form, and are more easily cured in Florida than in States to the north of it. I shall no doubt be met with the reply: "Look at the waxy complexions and gaunt forms of many Floridians, met with at some of the landings and depots." I admit the mild impeachment, and can attribute their cachectic condition to bad water, insufficient clothing, unsuitable and uncomfortable habitations, and the improper food they eat from childhood to the grave. In any other State but Florida they would be the victims of enlarged spleens, cardiac dilatation, chronic gastritis, tuberculosis, dropsical effusions, or albuminuria. But contrast the natives referred to with those who have comfortable homes, sufficient clothing, and who drink pure water and use good and nutritious food; or with Northern and Western people who have been in the State for years, and the latter will be found to be pictures of health. I admit that in Florida, as everywhere else, there are insalubrious localities, but they should be avoided by strangers. A majority of the cases of illness occurring among visitors in this State are referable to indulgence at table, drinking impure water, the inhalation of impure air, the American weakness of rushing hither and thither, occupation of unventilated rooms, and a ridiculous system of senseless drugging as a consequence of the advice given by physicians who are ignorant of the climate and its diseases.

From my observations in the United States and in foreign lands, and in hospital as well as in private practice, I have been forced to notice the infrequency of chronic disease and broken health in Florida. In my visits to various portions of this State I have met with many persons, old and young, who live from year to year on improper food, and who drink water from shallow holes, near marshes, and yet, singular to say (although such persons are somewhat anæmic), they do not present any manifest diseased condition. In cities, towns, villages, and rural districts, where residents are supplied with proper food and drink pure water, a case of chronic disease or broken

* Copeland's Dictionary of Medicine, vol. 1, p. 417.

health is seldom met with. And if we have a climate in which these conditions rarely
occur, are we not justified in concluding that it will exert a powerful influence in re-
storing the invalid to health? I have, at various times, visited many portions of the
State, and have been surprised to meet so many persons who have settled in it as invalids and have been restored to health or comparative comfort by the climate—a
large proportion of them having been sufferers from pulmonary diseases. And what
surprised me most, was the fact that none of their offspring manifested any constitu-
tional predisposition to pulmonary disease.

GEOGRAPHY AND TOPOGRAPHY. '

Florida is usually described as composed of East Florida, or that portion of the State between the Atlantic and the Suwannee River, comprising the whole of the peninsula; Middle Florida, extending from the
Suwannee to the Apalachicola; and West Florida, comprising the territory west of the last-named river. This division, suggested probably
by the existence of the distinctly-marked natural boundaries furnished
by the rivers named, may be well enough for the purposes of a general
description, but a different division suggests itself as better adapted to
the purpose of an agricultural description of different sections, whose
characteristic productions would be determined in the main by their
special climatic conditions. Accordingly, in attempting to give that
sort of practical information that would be serviceable to actual settlers,
and best enable them to make satisfactory locations, a different mode
of territorial division will be adopted, and for the purpose of properly
grouping the counties with special reference to those climatic conditions
which control and determine their vegetable productions, the State will
be included in the three divisions of Northern, Central, and Southern
Florida.

Northern Florida will be considered as constituted from all the lands
lying north of the parallel of 30° north latitude ; the territory included
between the parallels of 28° and 30° north latitude will be styled Central Florida; and all south of 28° will be considered as composing South
Florida.

Thus apportioned, a general account of each division will be given,
accompanied by such local descriptions of the different counties as will
convey a definite idea of the topography and characteristics peculiar to
each.

NORTHERN FLORIDA.

This division extends from the Atlantic Ocean on the east, to Perdido River on the west, a distance of 375 miles, and has an average
breadth of some 45 miles, and is composed of the counties of Escambia,
Santa Rosa, Walton, Washington, Holmes, Jackson, Calhoun, Franklin, Gadsden, Liberty, Leon, Wakulla, Jefferson, Madison, Taylor, Hamilton, Suwannee, Columbia, Baker, Bradford, Nassau, Duval, Clay, and
the northern part of Saint John's County.

The climate of this section as a whole may be said to be mild, verg-
ing upon warm. All extremes of temperature are essentially modified

by the prevalence of daily winds from the ocean or Gulf of Mexico. The eastern portion, probably from the influence of the Gulf Stream, has a milder and more agreeable climate than that west of the Suwannee, and in winter suffers less from the cold northers and northwesters that occasionally prevail. But through the whole section so equable is the climate that although ice may be formed two or three times in the course of a year, the thermometer very seldom falls below 35° in the winter, or rises above 90° in the summer. There are occasional frosts, but during four-fifths of the winter season the prevalent temperature is that of the mildest Indian summer at the North and West.

The surface of Northern Florida varies from the somewhat tame and monotonous levels of Eastern and Western Florida to the undulating and uneven lands of the middle portion, and gives a much wider field for selection than is commonly supposed, although extreme elevations are entirely wanting. Many strangers, who only make a personal inspection of the Saint John's region, and go away complaining of the tameness of the scenery, the lack of variety in the landscape, and the sluggish movements of the streams, would find in the valley of the Saint Mary's a deep and somewhat rapid stream inclosed between picturesque bluffs and high banks in the midst of a rich and fertile territory. The same is also true of the Suwannee, the Chipola, and other rivers.

From Hamilton on the east and Holmes on the west, the intermediate section is undulating, and in some parts quite broken; many portions of Jackson, Gadsden, and Leon Counties, in particular, bearing upon their surface a strong resemblance to the less hilly portions of Pennsylvania, New York, and New England, and thus is afforded in Northern Florida a variety of surface, consisting of sandy plains and "flat woods," and an alternation of hill and vale, from which the divers tastes of different individuals can be easily gratified.

The soils of Northern Florida are as varied as is the surface. To the east is found a light and sandy soil, with a subsoil varying in depth, of clay or marl. In the west the poorer soils are sandy, while the better are loamy in character. In the middle, or section commonly called "Middle Florida," strong clay soils are often to be met, as is especially the case in Jackson, Gadsden, and Leon Counties.

It is undeniable that here, as in the State generally, there is a good deal of light and poor soil; but it is equally true that as rich and productive lands exist in Northern Florida, and when considered with reference to the value of the staple crops, as productive and valuable lands as can be found anywhere. The first year's crop of cotton and corn has in more than one instance repaid the purchase price of a plantation.

From the broken and uneven surface of the middle counties on the north, and from the comparative coolness of the climate, this division of the State is better adapted than either of the others to what is understood by ordinary "farming," as contradistinguished from "planting," so called. Hence there is a greater diversity of the crops usually raised

3290——2

in the Northern and Middle States than can ordinarily be found in the other divisions. Here, instead of depending mainly upon the avails of a single crop, as is usual with cotton, rice, and sugar planters, there is more of what is understood by the term of farming, and each cultivator aims at raising his own supplies as far as practicable; and cotton, corn, cane, wheat, rye, oats, hay, potatoes, and tobacco are often, and indeed commonly, made by each individual farmer.

The staple crops may be said to be corn, cotton, tobacco, rice, cane, Irish and sweet potatoes, and oats. The principal fruits are peaches, figs, grapes, oranges, though many others are raised. The apple and pear do not generally flourish, except along the Saint Mary's River (which is one of the best fruit regions in the whole South) and in the strong clay soils of Jackson, Gadsden, and Leon Counties. The peach and fig thrive everywhere and mature several weeks earlier than in the States lying north. The orange is grown throughout this division, the crop increasing in security as you go south; but very fine oranges are raised in the northern counties, although, when young, they should have some protection. No better oranges are raised in Florida than those produced in Liberty, Calhoun, Wakulla, and Washington Counties, and the world can show no better oranges than the best raised in this State.

This whole division is remarkably well watered. In addition to the numerous rivers—the Perdido, Black Water, Escambia, Econfina, Apalachicola, Chipola, Ocklockonee, Ancilla, Suwannee, Saint John's, Saint Mary's, and Nassau—and the multitude of smaller streams, nearly the whole region is abundantly supplied with springs, while good water is easily obtained in wells of little expense.

The timber of the region is abundant. The supplies of pine and cypress are apparently inexhaustible; while hickory, oak, ash, cedar, magnolia, and red bay are found everywhere. Game and fish are found everywhere in good supply. On the coasts, oysters and turtle abound. They are both abundant and good on the east coast; but the oysters of Saint Andrew's Bay, on the west, are not surpassed for size and flavor, and are exceedingly abundant.

So much will suffice for a general description of Northern Florida as a whole, and for more particular information reference is made to the local descriptions of each county in this subdivision, arranged in alphabetical order.

BRADFORD COUNTY.

Bounded north by Baker County, east by Clay, south by Alachua, and west by Columbia County. Area, about 600 square miles. The surface mostly level, but sufficiently high and undulating for cultivation. The soil varies from light to strong pine land, and is covered, where not improved, with a heavy growth of pine timber. This timber and naval stores are largely exported. The Atlantic, Gulf and West India Transit Railroad, which runs from Fernandina, on the Atlantic, to Cedar Keys, on the Gulf of Mexico, runs southwest across the eastern border of the county. This is a progressive county, and has a thrifty population of old and new settlers. All the usual crops do well, and the orange groves look as well as in any section; market gardening is also profitable here.

Lake Butler is the county seat, though Starke is the largest, both in population and business, and is situated directly on the line of the railroad.

BAKER COUNTY.

Bounded north by Georgia, east by Nassau and Duval Counties, south by Bradford, and west by Columbia County. Area, about 500 square miles. Its topography, soil, and general characteristics are similar to Bradford County. Mostly level, heavily timbered; soil, where cultivated, productive. The Central Railroad runs through the county from east to west, furnishing easy transportation to Apalachicola River on the west, to Jacksonville and Fernandina on the east, and connecting with the railroad system north. Sanderson, on the line of the railroad, is the county seat. Many small farmers are settling in this county.

COLUMBIA COUNTY.

Bounded north by Georgia, east by Baker and Bradford Counties, south by Alachua, and west by Suwannee and Hamilton Counties. Area, about 864 square miles. Its soil includes pretty much every variety found in Florida. The surface is generally level; the southern portion well timbered: the western part, high rolling pine lands of good quality; very little waste land unfit for cultivation. There are twelve lakes of moderate size, Alligator Lake being the largest; all abound in good fish. Muck, marl, limestone, sandstone, and clay suitable for bricks are found. Several streams afford good mill sites, and at Suwannee Shoals, on border of Hamilton County, there is sufficient water for large manufacturing establishments.

The railroad from Jacksonville runs through the county from east to west, with a depot at Lake City. Lake City, the county seat, is a neat place, surrounded by lakes; is the center of a well-settled agricultural country, and does a large commercial business. Besides cotton, cane, rice, tobacco, corn and root crops, raising vegetables for shipment to North and West is becoming a large industry. Orange and grape culture is receiving special attention, with best results. Some of the largest vineyards in the State are in this county.

CLAY COUNTY.

Bounded north by Duval County; east by Saint John's River, which separates it from Saint John's County; south by Putnam, and west by Bradford County. Area, about 425 square miles. The county is well watered, sufficiently high and uneven to afford water-power on several streams. Black Creek traverses the county, and is navigable for river steamers to Middleburg, the center of the county. The Atlantic, Gulf and West India Transit Railroad crosses the northwestern township of the county, about twelve miles from the head of navigation on Black Creek, so the county has excellent facilities to reach markets by water or rail.   There are several fine lakes in the southwestern portion of the county, which afford, at all seasons, an abundance of food fish.   Lake Kingsley is the largest, in the near vicinity of which, and in the section lying between the lake and the railroad, settlements and improvements are being rapidly made.   Most of the soil of this county produces well all the staples of the-country; and the usual vegetables and varieties of fruit.   Bordering the many streams and lakes there are rich hammocks. The land, where not opened, is well covered with pine.   Marl beds of large extent are found, and on Black Creek fine stone for building pur- poses.   Middleburg, formerly the county seat, a town once of consid- erable importance, at head of navigation, was formerly the place of trans- shipment to and from the interior.   The building of the railroad from Fernandina has diverted this.   The county seat, Green Cove Springs, on the Saint John's, is a thriving place, and a great resort both for win- ter travelers and others who seek benefit from the sulphur spring, which is large.

CALHOUN COUNTY.

Bounded on the north by Jackson County, east by the Apalachicola River and Franklin County, south by the Gulf of Mexico, which, with Washington County, forms the western boundary.   Area, 670,000 acres. Lands in the northern part are elevated and rolling; in the southern por- tion level, and in some places low.   The Apalachicola River is navigable for large steamers, and the Chipola River, which nearly bisects the county from north to south, is navigable a portion of the year.   Other streams abound and afford ample water-power, which is used, whenever desired, to advantage.   Chipola Lake, 16 miles long and from 1 to 3 miles wide, is situated near the center of the county, and is full of fish of many kinds; the forests abound in game.   Very extensive beds of marl, some quarries of stone, and clay suitable for brick are found in this county. Cotton, sugar-cane, corn, and peanuts, are the principal crops raised, as also vegetables and root crops.   Orange culture is rapidly extending, and successfully.   Stock-raising is carried on to some extent, and profit- ably.

DUVAL COUNTY.

Bounded north by Nassau County, east by the Atlantic Ocean, south by Saint John's and Clay Counties, west by Baker and Nassau.   It has an area of about 860 square miles, embracing the mouth of the mag- nificent Saint John's River, the natural outlet of nearly a thousand miles of inland navigable waters.   While the lands as a whole are not as rich in an agricultural sense as some other sections, yet there are to be found large and small tracts of rich hammock.   Most of the land, however, is light, but the modifying influence of the waters of the ocean and the broad Saint John's and other streams are favorable for crops, and especially for vegetables and fruit. Add to this the commercial facilities of river navigation, outward to sea and interior, the railroads connecting north, west, and south with the great through lines, and Duval County offers the very best advantages for general Southern crops, and particularly for large and small fruit growing and market gardening, which is rapidly extending. The light lands respond quickly to fertilizers, and marl, shells, and muck are found within easy distance.

Jacksonville, the county seat, is in the center of the county, and is the largest and most thriving city in East Florida, and in the very near future may rank in commercial importance with Savannah and Charleston. It is healthy, has many fine hotels, a complete system of water supply; thorough sewerage, rigid sanitary and police regulations, and is every way progressive. The Jacksonville, Pensacola and Mobile Railroad, the Savannah and Florida, and the Fernandina Railroads intersect the Saint John's River at this point, and the Jacksonville, Saint Augustine and Halifax River Railway is in process of construction. Arrangements have also been made for building a railroad from this city to Key West, touching at Palatka, Titusville, Tampa, and Punta Rassa. Steamers ply to Savannah and Charleston, connecting at those points with steamers to the principal Northern ports. Lines of schooners run regularly to New York.

ESCAMBIA COUNTY.

This county forms the west end of the State. Bounded north and west by Alabama, east by Santa Rosa County, and south by Gulf of Mexico. Perdido Bay and River separate it from Alabama on the west, Escambia River and Bay from Santa Rosa County on the east. That portion of the county bordering the Gulf is level, with a light soil, covered with pine; where this has been cut off, oak, hickory, bay, magnolia, and other hard woods have succeeded. The northern part, being over two-thirds, is uneven and hilly, and clay subsoil is near the surface, occasionally cropping out. The area of the county is about 600 square miles.

Pensacola is the county seat, and one of the most beautifully located and important cities in the State. Pensacola Bay has no equal in the Southern States, and in capacity, depth of water, and safety is not excelled by any Northern port. There is a large and well-equipped United States navy-yard, dry-dock, and coal-depot, as also Fort Barrancas, Fort Pickens, and the old Fort McRae. The recent extension of railroads to Pensacola, connecting it with the great through lines west, north, and east, will make it a great shipping port for products of field, mines, and manufactures. Escambia River is navigable for steamboats for twenty-five miles, and the Perdido for same distance. A railroad connects Perdido Bay with Pensacola Bay.

FRANKLIN COUNTY.

Bounded on the north by Liberty and Wakulla, east and south by the Gulf of Mexico, and west by Calhoun County. It is divided by the Apalachicola River, and includes Apalachicola Bay, Saint George's Sound, and the adjacent islands. It contains about 600 square miles, and was formerly one of the most thriving and important counties of the State. Apalachicola, the county seat, was formerly a place of large commerce; the lines of railroad from Atlantic cities west have almost entirely diverted the trade, and from being one of the largest cotton ports of the South, it has become only the port for a limited area of country. But with a fair port on the Gulf, and steamboat navigation reaching into Georgia and Alabama, by the Apalachicola, Chattahoochee, and Flint Rivers, there is a good prospect of its future growth as the country becomes settled. Many portions are rich, alluvial bottoms, very productive. All the Southern staples are successfully cultivated, and the orange and semi-tropical fruits fully equal, in growth, quality, and quantity, those of other sections. The bays and waters of the Gulf afford superior fish and oysters, and yield abundantly.

GADSDEN COUNTY.

Bounded north by Georgia; east by Leon, from which it is separated by the Ocklockonnee River; south by Leon and Liberty; west by Jackson County, from which it is separated by the Chattahoochee River. It contains an area of over 450 square miles. The surface is uneven, elevated, and presents a strong contrast with the more level lands on the Atlantic and Gulf coasts, the topography and soil in many portions resembling the northern parts of Virginia. It is one of the best-watered portions of the State; clear running streams and springs of freestone water are met with at short intervals, in every direction, which afford water-power for manufacturing. The soil is mostly based on red clay, lying from a few inches to two feet beneath the surface; the lands being rich, productive, and durable, are thus adapted for the growth of grain and forage crops, also cotton and cane. Previous to the war, this county was noted for its production of superior tobacco, which equaled Cuba tobacco in quality and price. The export previous to 1860 was 400 boxes of 400 pounds each of tobacco, annually. It is among the richest agricultural counties in the State, and has little waste land, and a larger proportion under cultivation than most others.

Quincy, the county seat, is a beautiful town, its location on high elevation affording fine views of the surrounding country. The Jacksonville and Mobile Railroad crosses the county from east to west to the Apalachicola River, thus affording good facilities for transportation to the North and West as well as to Eastern and Southern ports. Beds of marl are found in this county, as also clay suitable for brick.

HAMILTON COUNTY.

Bounded north by Georgia, east by Columbia, south by Suwannee, and west by Madison County. Area about 400 square miles. The Suwannee River forms its southern and eastern boundary, the Withlacoochee River its western, the Alapaha River dividing in nearly in the center. The Savannah and Gulf Railroad crosses from south to north, nearly in the center of the county. The general surface is rolling, soil light in some portions, with fine hammocks near streams. Jasper is the county seat.

HOLMES COUNTY.

Bounded north by Alabama; east by Jackson, from which it is separated by Holmes Creek; south by Washington and Walton; west by Walton County. Area over 500 square miles. The Choctawhatchie River runs south through the center of the county, affording steamboat communication with the Choctawhatchie Bay and the Gulf. Stock-raising, cotton-growing, and farming the principal business; sugar-cane, corn, potatoes, and other food crops raised for sale and home consumption. The land is mostly level, good pine lands, well timbered, varied by rich hammocks. The great need of this and adjoining counties is railroad communication. The Pensacola and Mobile Railroad will pass through this county at or near Cerro Gordo, the county seat, which is pleasantly situated on the high banks of the Choctawhatchie. A railroad is in process of rapid construction that will connect the county with Pensacola and Jacksonville.

JACKSON COUNTY.

Bounded on the north by Alabama; east by Decatur County, Georgia, and Gadsden County, Florida, from which it is separated by the Chattahoochee and Apalachicola Rivers; south by Calhoun and Washington Counties, and west by Washington and Holmes Counties; has an area of 1,000 square miles. It is considered as one of the richest agricultural districts of the State. Lands are rich, undulating hammock; soil composed of clay, loam, and lime, in various proportions, and pine lands of good quality of soil. The Chipola River, rising in Alabama, flows south nearly through the county, emptying into the Apalachicola; is navigable for boats of moderate draft. At a small expense the river could be made navigable for steamboats. The Apalachicola and Chattahoochee Rivers, on the eastern boundary, afford transportation to markets. The county exports largely cotton and other agricultural products.

Marianna, the county seat, is a beautiful place, doing a large business; is located on the Chipola River, the lower valley of which is well adapted to orange-growing, as also other fruits; soil rich and remarkably exempt from frost. The extension of the Jacksonville, Pensacola

and Mobile Railroad, now rapidly progressing, will afford facilities for
communication east and west, which cannot fail to render the county
attractive for immigrants and capitalists.

JEFFERSON COUNTY.

Bounded north by Georgia; east by Madison and Taylor, from which
it is separated by the Aucilla River; south by Taylor County and the
Gulf of Mexico; west by Wakulla and Leon. Has an area of about 550
square miles. It occupies a central portion in the tier of counties
known as Middle Florida, and offers many and substantial inducements
to immigrants, especially to those who seek homes where they can enjoy
all the benefits of civilization and the facilities for easy and cheap communication with markets. The face of the county, from the Georgia
line south for about 20 miles, is beautifully undulating, intersected
throughout with small streams fed by springs, and dotted here and
there with beautiful lakes, prominent among which is the Miccosukie,
which extends into Leon County and is 15 miles long by from 1 to 4
miles wide. The southern portion of the county is mostly flat woods.
The soil is varied—in the upper and middle are stiff, red lands, with clay
subsoil; on the Aucilla River and bordering on the flat woods are rich
hammocks. It has a larger proportion of cultivated lands than other
counties, and is among the largest cotton-producing counties. With
Madison, Leon, and Gadsden it constitutes what is known as the rich
agricultural district of Northern Florida.

The Jacksonville, Pensacola and Mobile Railroad crosses the county
near the center, with a branch to Monticello, the county seat, one of the
most healthy and delightful villages in the State.

LEON COUNTY.

Bounded north by Georgia, east by Jefferson County, south by Wakulla County, and west by Liberty and Gadsden Counties, from which it
is separated by the Ocklockonnee River. It contains an area of about
700 square miles. The surface of the county, like that of the adjoining
counties, which constitute what is known as Middle Florida, is varied;
the northern portion uneven, the southern more level and interspersed
throughout with clear-water lakes, among which are Lafayette, Jackson,
Iamonia, Bradford, and the Miccosukie, extending from Jefferson County,
all abounding in fish. The soil is as varied as the surface. In the northern half of the county it is rich loam, based upon red clay, very productive. In the southern portion the soil is lighter, the clay lying deeper
and of a pale yellow color. Leon is the center of the rich agricultural
counties of Northern Florida, and no district of the same extent in the
country can offer superior inducements to cultivators of the soil. Short
staple cotton has been the principal source of reliance, but wheat, corn,
rice, rye, oats, sugar-cane, tobacco, and all the diversified products of a rich
agricultural district are successfully cultivated, and all kinds of stock

raised with profit. Whether we consider its unexceptionable climate, the beauty of its undulating surface, the variety, abundance, and value of its timber, the fertility of its soil, with its adaptability to such a vast catalogue of crops, its accessibility to markets, its abundance of good, pure water, its general healthfulness, the ease with which the soil is cultivated, the intelligence and character of its people, the number of its laboring population, or the cheapness of its lands, no portion of the State or the country can offer superior inducements to immigrants.

The beautiful city of Tallahassee, the county seat and capital of the State, lies near the center of the county. The Jacksonville, Pensacola and Mobile Railroad crosses the county from east to west, and the Tallahassee Railroad, from the port of Saint Mark's, on the Gulf, intersects it at the capital. The name Tallahassee, signifying *beautiful land*, was applied by the Indians to this region of country, and was properly appropriated to designate the capital at the time of its location.

LIBERTY COUNTY.

Bounded north by Gadsden; east by Leon and Wakulla Counties, from which it is separated by the Ocklockonnee River; south by Franklin; west by Calhoun, from which it is separated by the Apalachicola River. It contains an area of about 850 square miles. It is sparsely populated, and little of its area is cultivated. Its characteristics are the same as Calhoun and Franklin. Cattle-raising is the principal business, but the ordinary staples of the State are successfully cultivated; orange culture is receiving attention, and fine groves attest the success of this important product. Bristol is the county seat.

MADISON COUNTY.

Bounded north by Georgia; east by Hamilton and Suwannee Counties from which it is separated by the Suwannee River; south by Lafayette and Taylor, and west by Jefferson, from which the Aucilla River separates it. It contains 750 square miles, and forms a portion of the rich agricultural district of Middle Florida. The lands are generally undulating, and some portions even hilly; a small part of the southeastern portion is flat. The western half is very fertile, the eastern generally pine lands of fair quality and interspersed with lakes and ponds. The better lands are generally underlaid with clay, and the soil rich and generous. Long and short staple cotton have formed the chief product for exportation, and as high as 12,000 bales have been produced in a year. Now, while cotton continues the principal staple, the products are becoming more diversified, and grass, grain, sugar cane, and vegetables are receiving more attention and are found remunerative, while stock-growing and fruit culture are successfully prosecuted. A larger proportion of the lands of Madison County are under cultivation than of any other county.

Madison, the county seat, is a thriving place near the center of the county, on the Jacksonville, Pensacola and Mobile Railroad, which crosses the county from east to west. One of the largest lumber manufacturing establishments in the State, employing a capital of over $300,000, is situated in the eastern portion of the county; $30,000 are invested in grist-mills.

NASSAU COUNTY

Occupies the northeast corner of the State, and is bounded north and west by the Saint Mary's River, which separates it from Georgia, east by the Atlantic Ocean and Duval County, and south by Duval. It contains about 600 square miles, including Amelia Island, upon which the city of Fernandina, the county seat, is located. The soil of Nassau County varies from the light mulatto soils of the coast, through all the intermediate gradations, to the stiff clays and marls in the lowlands of the rivers, and its range of productions is as varied as the soil. On Amelia Island, the edge of the mainland, and scattered along her rivers, are soils of calcareous sand, that are adapted for the finest qualities of long staple cotton, and the culture of the peach, grape, olive, and orange, while the fresh marsh and black rush lands attached to them are especially suitable for gardening. These lands are easily reclaimed, rich, moist, and close to shipping opportunities, so that the shipping of early vegetables to Northern markets forms a profitable industry. The clay bluffs along the Saint Mary's River, and the so-called sand hills in the northwestern corner of the county, form a third distinct body of agricultural lands. The balance of the lands of the county are pine barrens, mostly sandy, and interspersed with numerous "bay-galls," cypress ponds, and savannas. The harbor of Fernandina is the northern terminus of the Atlantic, Gulf and West India Transit Railroad, from Cedar Keys, and is one of the best harbors for sea-going vessels of deep draught south of Norfolk, admitting of the safe anchorage of several hundred vessels at once, and with an entrance easy of access, and giving from 14 to 20 feet of water.

SAINT JOHN'S COUNTY.

Bounded north by Duval, east by the Atlantic Ocean, south by Volusia, and west by Putnam and Clay Counties, from which it is separated by the Saint John's River. It contains 970 square miles. The general surface is level, and the land is not of the first quality, being mostly flat pine woods and palmetto scrub, with some hammock; but its location, between the Saint John's River and the Atlantic Ocean, renders it more exempt from frost and better adapted to fruit culture than more interior counties in the same latitude. Orange culture and fruit and market gardening are now commanding attention, while stock-growing, corn, rice, sugar cane, &c., are profitable branches of agricultural industry. The Matanzas and North Rivers lie parallel with the coast on the east, and the Saint John's River and Lake Crescent on the western border.

Saint Augustine, the oldest city on the continent, rich in historic interest, and once famous for its orange groves, from which for nearly a century the nobles and grandees of Spain derived their best supply, is the county seat, and a port of entry for sea-going vessels, and is connected with the Saint John's River by railway to Tocoi, and a railway direct to Jacksonville is in process of construction. It is proverbial for its healthy and delightful climate, and is a popular resort, both summer and winter, for visitors seeking health and recreation.

SUWANNEE COUNTY.

Bounded on the north by Hamilton County, east by Columbia, south by Alachua and La Fayette, and west by La Fayette and Madison, from which, with Hamilton on the north, it is separated by the Suwannee River. Its area is about 700 square miles. This county occupies nearly a central position, from east to west, in the State, and the Suwannee and Santa Fé Rivers form its boundary on three sides, a distance of over 100 miles. These streams are navigable for steamboats to the southeastern part of the county. The general topography is rolling. The soil is sandy, in some parts mixed with a clay subsoil. Beds of marl, shell, and white clay fine enough for pottery. Limestone and sandstone abound, the latter white as marble, and, when first exposed, so soft that it may be cut into any desirable form, and becomes hard with exposure. Lumbering and naval stores form the leading industry, as the timber is very fine. The Jacksonville, Pensacola and Mobile Railroad crosses the county from east to west, and is intersected from the north by the Savannah Railway at Live Oak, the county seat, which is now in process of extension south, with Tampa and Charlotte Harbor as the objective points.

SANTA ROSA COUNTY.

Bounded north by Alabama, east by Walton County, south by the Gulf of Mexico, and west by Escambia, and contains about 1,600 square miles of territory. The surface and soil and the natural productions are very nearly like those of Escambia, which joins it on the west. Lumbering is the principal business, and agriculture has received little attention. The country is well watered, the Escambia River and Bay form its western boundary, and Pensacola Bay and Santa Rosa Sound lie upon its southern, while the Yellow, Black Water, and Clear Water Rivers and various creeks divide the interior of the county and discharge their waters into Pensacola Bay. Milton, located at the head of the bay and at the mouth of the Black Water, is the chief town and county seat. A large foreign export trade in lumber and timber has long been conducted from this port.

TAYLOR COUNTY.

Bounded north by Madison County, east by La Fayette, south by the Gulf of Mexico, west by the Gulf and Jefferson County; and has an area of 1,300 square miles. The Aucilla River enters the Gulf on its western boundary, and the Isteenhatchie on the eastern, while the Econfina, Fin-holloway, and Warrior lie intermediate. There are several sulphur, iron, and chalybeate springs. The surface is generally level, the lands are pine and hammock, and toward the Gulf coast are comparatively poor. The streams abound in fine fish, the coast in oysters, and the forests in game. It is a fine range for cattle, and stock-growing is the leading business; though cotton, corn, sugar cane, and tobacco for home use are produced by the farmers. Perry is the county seat.

WAKULLA COUNTY.

Bounded on the north by Leon, east by Jefferson, south by the Gulf of Mexico, and west by Franklin and Liberty Counties, from which it is separated by the Ocklockonnee River. It has an area of about 650 square miles. The surface is generally level, though sufficiently undu-lating for drainage. The lands vary from light pine to the richest ham-mock, and are intersected by streams, the most important of which are the Saint Mark's, Wakulla, and Sopchoppy. There are numerous springs—sulphur, chalybeate, and pure water. The sulphur springs at Newport, in the eastern part of the county, were formerly a popular re-sort for invalids, and the famous Wakulla springs, whose transparent waters create the sensation, while floating on its surface, of being sus-pended in the air, forms one of the most wonderful and attractive feat-ures. The port of Saint Mark's, at the mouth of the river of that name and the terminus of the Tallahassee Railroad, was formerly a place of considerable commercial importance; the construction of the various lines of railway from the Atlantic ports has diverted this trade. The streams abound in fish, and the coast in oysters, and with the facilities for communication with the markets of the world there are abundant inducements for settlement and cultivation. Stock-growing and agri-culture are the leading industries.

Crawfordsville is the county seat, and is near the center of the county, and in a fertile and productive portion. The extension of the railway, now in progress, will soon afford communication east and west with the railway system of the country.

WALTON COUNTY.

Bounded north by Alabama and Holmes County, east by Holmes and Washington, south by the Gulf of Mexico, and west by Santa Rosa County. It embraces an area of about 1,800 square miles. The county was first settled in 1823, by a colony of Scotch families, who located in Uchee Valley, and whose decendants still possess the land. The lands of Walton County are principally pine; along the eastern boundary much of the soil is light, but there are exceptions, notably on the Choc-tawhatchie, where there is a tract 15 miles long by 4 miles wide of ex-ceeding fertility. In the Uchee Valley is another tract of rich land, with clay subsoil, of about the same extent. Along the eastern and southern boundaries water communication with the Gulf of Mexico is furnished by the Choctawhatchie River and Bay, both navigable for steamboats, while the interior is watered by numerous creeks and runs, some finding their way into the Choctawhatchie, and others passing west into Pensacola Bay. The Jacksonville and Pensacola Railroad will cross this county.

CENTRAL FLORIDA.

This division is made up of the territory lying between the parallels of 28° and 30° N. latitude, and is composed of the counties of La Fayette, Alachua, Levy, Marion, Putnam, Volusia, Orange, Sumter, Hernando, and the southern portion of Taylor, Clay, and Saint John's Counties.

The surface of this division is less broken, and, as a whole, more level than Northern Florida. It has more of savanna and marsh, and is bountifully supplied with water, having the Stinhatchie, Suwannee, Santa Fé, Withlacoochee, Crystal, Hillsborough, Acklawaha, and Saint John's Rivers, and is profusely studded with ponds, lakes, and smaller streams. The climate is very perceptibly milder, not only from its more southern geographical position, but the narrowness of the peninsula here, giving an average breadth between the ocean and the Gulf of only about ninety miles, exposes it to the daily sweep of the winds from either side, and by this means the extremes of both heat and cold are very essentially modified and ameliorated. The exposure to daily winds from each side increases, also, the rain supply, so that this division has more frequent and abundant rains, and suffers less from drought, than the northern division.

The soils of Central Florida are similar to those of Northern Florida, with a large proportion of hammock and savanna, and are perhaps of better quality, as a whole. Levy, Hernando, Alachua, Marion, and Sumter Counties, taken together, form a body of land that for productive capacity is not excelled by any similar body in the United States.

The staple crops are similar to those of Northern Florida, but the peculiar adaptability of this division to the cultivation of the sugar cane and all semi-tropical fruits has caused cane to advance rapidly of late in the estimation of farmers, and within a few years it will probably become the leading agricultural production. The sugar cane in this division ratoons for six or eight years in succession without protection, and often attains a height of from 10 to 15 feet, even when grown for a number of years on the same land without manure.

Particular attention is asked to the statistical return of crops in Hernando County, which is appended, and which, with other facts given, fully sustains the assertion that Central Florida is the best cane region in the United States, and probably in the world.

The entire division is the natural habitat of the whole citron tribe; numerous groves of the wild orange have been found and still occasionally appear, and, as would naturally be anticipated, the orange, lemon, and lime are natural and very prolific and profitable crops. The peach and the fig thrive everywhere; the guava and the banana do well without protection; and the pine-apple is cultivated, although it does not flourish as in South Florida. Irish and sweet potatoes, melons, and all kinds of garden vegetables are cultivated with great success, and can be brought to maturity at almost any season, at the option of the cultivator.

ALACHUA COUNTY.

Bounded north by Suwannee, Columbia, and Bradford Counties, from which it is separated by the Santa Fé River; east by Clay and Putnam; south by Marion and Levy; and west by La Fayette County, from which it is separated by the Suwannee River. It has an area of over 1,300 square miles, and embraces almost every variety of lands found in the State, from the richest hammock, high rolling pine, hickory, and oak, to the more level, heavy timbered pine lands. Its elevation above the sea is from 50 to 250 feet; it has numerous lakes and streams, which afford good water-power and abound in excellent fish. Lake Santa Fé, in the northeast portion of the county, is believed to be the highest body of water in the eastern portion of the State, being on the ridge from which waters flow to the Atlantic Ocean on the east and to the Gulf of Mexico on the west.

The Atlantic, Gulf and West India Transit Railroad runs directly through the county, from the northeast to the southwest, entering the county near Santa Fé Lake. The Peninsular Railroad intersects it at Waldo, a growing and thriving town in the northeastern portion of this county, and runs nearly due south to Orange Lake, some twenty miles, and is being extended to Ocala and thence to Tampa Bay, on the Gulf. A canal is nearly completed from Waldo, connecting with railroad, and also connecting Lakes Alto, Santa Fé, and smaller lakes, opening up a large area of excellent lands to easy access.

Gainesville, the county seat, is one of the most progressive towns in the State, both in population and business. The population of the county is increasing from year to year, and the agricultural and horticultural resources are developing more and more each season.

The fruit and vegetable industry, only as yet in its infancy, has already assumed large proportions; 450,000 packages alone were shipped over the railroad the past season. The staple crops are, long-staple and upland cotton, sugar cane, rice, corn, root crops, vegetables of all kinds, oranges and other semi-tropical fruits. This county was early selected by the pioneer settlers as one of the best in the State, and time has proved the wisdom of their choice.. From the northwest to the southeast a peculiar limestone formation is found, the crust in many places having, from some cause, sunk down; these depressions are generally more or less full of water, and connected by subterranean waters; these sinks are oval and conical downwards, and from 10 to 40 feet deep.

HERNANDO COUNTY.

Bounded north by Marion and Levy; east by Sumter, being separated from these counties by the Withlacoochee River; south by Hillsborough County; and west by the Gulf of Mexico. Its area, over 1,600 square miles, fronts on the Gulf 60 miles, extending from the mouth of the Withlacoochee south to Anclote River, embracing Crystal, Homosassa, Cheseehowiska, Wekiwachie, Pithlochascotee, and Anclote Rivers, which vessels of ordinary draft can enter.

No county in the State has a more varied topography, or greater advantages for the successful prosecution of agricultural and horticultural pursuits, or is so attractive for a residence.

Lands high and rolling, like the red hills of Northern Georgia; high, smooth tracts of pine lands; extensive hammocks of the richest soil; frequent marl beds; limestone; large springs of the purest water; lakes and rivers abounding in fish; a long coast with frequent harbors; the bays and Gulf always afford fish, oysters, and sponge; a climate and soil adapted to cultivation of cotton, cane, rice, tobacco, corn, oats, grass, and vegetables, having peculiar advantages for growing the olive, the different varieties of the citrus, the pine-apple, guava, banana, and all the semi-tropical fruits. Transportation is year by year becoming more . rapid and cheap, and access to and from markets easier. Immigration, enterprise, and industry will make it one of the most prosperous and desirable portions of the South. The county seat is Brooksville.'

LEVY COUNTY.

Bounded north by Alachua, east by Marion, south by Hernando and the Gulf of Mexico, and west by the Gulf and La Fayette County, from which the Suwannee River separates it. It has an area of over 1,000 square miles. The surface is generally level, being mostly flat pine wood land. The Gulf hammock, a tract of land of great fertility, of some 100,000 acres, capable of producing sugar cane equal to Louisana bottoms, occupies the southern portion of the county. The Suwannee River enters the Gulf on the western boundary, the Withlacoochee on the southern, with the Wacasassa about midway between. The Atlantic, Gulf and West India Transit Railroad runs from northeast to southwest through the county, near its center, and intersects the Gulf at the harbor of Cedar Keys, where vessels find entrance, and freight and passengers are transferred from the Gulf steamers to the railroad, thus affording enlarged facilities for direct communication with the markets of the North and the ports of the Gulf. The county possesses peculiar advantages for the production of sugar cane and rice, besides the ordinary products of long-staple cotton, vegetables, semi-tropical fruits; and stock-growing forms a sure reliance for revenue. The waters on the coast abound in fish, oysters, and turtle, which are largely gathered for export to the interior.

Bronson, the county seat, is on the railroad, the centre of a well-settled portion of the county.

LA FAYETTE COUNTY.

Bounded north by Suwannee; east by Suwannee, Alachua, and Levy, from all of which it is separated by the Suwannee River; south by the Gulf of Mexico ; and west by Taylor County. Contains an area of over 1,200 square miles. The land is principally heavy timbered pine lands, with many large tracts of hammock, a portion with a strong clay foundation and productive. The principal business is stock-growing and lumbering, but it is eligible for agriculture and fruit-growing, and the Suwannee, which skirts its eastern boundary, is navigable for steamers to New Troy, the county seat.

MARION COUNTY.

Bounded north by Alachua and Putman, east by Putman, Volusia, and Orange, south by Sumter and Hernando, and west by Levy County. Containing an area of 1,000 square miles. It is one of the largest, most fertile, productive counties of the State, especially in Sea Island cotton and sugar cane. The lands are generally elevated and undulating, and drain both to the Ocean and Gulf. There is very little poor and un- available lands, the most being rich and productive. The pine lands are almost uniformly good, underlaid with clay, marl, limestone. The ham- mocks are extensive and very rich, and will equal the best lands of the Mississippi in producing. Sandstone for building purposes is in great abundance. The Ocklawaha River, a tributary of the Saint John's, and navigated by steamers daily, runs north across the center of the county.

The celebrated Silver Spring forms a basin of two or three acres in extent near the center of the county; it pours forth a volume of water from one to two hundred feet wide, discharging into the Ocklawaha. Blue Spring, almost as remarkable, and not much inferior in size, lies in the northwestern portion of the county, and sends forth a stream of clear blue water into the Withlacoochee River, some twenty miles from the Gulf. Sulphur springs are numerous ; the most noted is known as Orange Spring, in the northeastern portion of the county, which was formerly a great resort for invalids. Orange Lake, celebrated for the large orange groves on its borders, which are the most extensive of any in the State, occupying an area of over 1,000 acres, lies in the northern portion of the county, and is now connected by the Peninsular Railroad with the Atlantic, Gulf and West India Transit Railway at Waldo.

Lakes Churchill and Bryant, in the eastern, and the beautiful Lake Weir, in the southern part of the county, are the most prominent and attractive of the inland water of the county.

Ocala, the county seat, situated six miles from Silver Spring, is a growing business town. The Peninsular Railroad is completed to this place, from which it will be extended southward to Tampa and Charlotte

Harbor, on the Gulf of Mexico. The Florida Southern Railway from Palatka, on the Saint John's, and Gainesville, on the Fernandina Railway, is finished to this place, and is to be extended south to Tampa and Charlotte Harbor. The government lands as well as State lands are being rapidly taken up by homestead and purchase. No part of the State, or, in fact, the South, offers greater inducements for permanent location.

ORANGE COUNTY.

Bounded north and east by Volusia County, which is separated from it by the Saint John's River, south by Brevard and Polk, and west by Polk, Sumter, and Marion; with an area of 2,300 square miles. The county is generally high, rolling pine land, interspersed with clear-water lakes, bays, and hammocks. The rolling pine lands are of good quality and heavily timbered; soil dark gray loam, with sand on the surface, based upon yellow sandy loam, with a substratum of clay and marl. Portions are flat pine woods of less value. Some of the prominent lakes are Monroe, Jesup, Harney, Eustis, Apopka, Dora, Maitland, Butler, and Tohopekaliga. These lakes are from 3 to 50 square miles in extent. There are innumerable smaller lakes, with areas of from 10 to 1,000 acres. The shores are generally abrupt, rising in some cases to 70 feet above the water. Fish and game abound. Stock-growing has been the predominant industry until later years, with cotton, corn, and cane; but now fruit culture is absorbing general attention, and the orange, lemon, lime, citron, guava, pineapple, and banana, and every variety of Southern fruit, are extensively cultivated. No county in the State has increased in population and improvement so rapidly during the last ten years as Orange, and large accessions from the Northern and Western States, of refined, cultured, and wealthy citizens, are constantly being made. A railroad from Sanford, on Lake Monroe, the head of the larger class of steamboat navigation, to Orlando, the county seat, has been constructed, and is in process of construction south through the county, and to Tampa and Charlotte Harbor. The Saint John's and Lake Eustis Railway, from Astor, on the Saint John's, to Fort Mason, on Lake Eustis, has also been completed, an extension of which to Leesburg will probably be made. The industry, energy, and progressive spirit manifested in this county is of the character manifested in the North and Northwest, and cannot fail of ultimate success.

Population, 1870, 2,195; 1880, 6,190—white, 5,494; black, 696. Number acres land tilled, 13,166. Farm values, $3,381,410; farm implements and machinery, $52,040.

PUTNAM COUNTY.

Bounded on the north by Clay County, on the east by Saint John's, on the south by Volusia and Marion, and on the west by Alachua and Clay, and contains an area of over 800 square miles. The Saint John's River runs through the county for 30 miles, and skirts it on the east for re-

3290——3

mainder, affording means unexcelled for transportation. Lake George, a body of water about 10 miles wide by 20 long, is on its southern boundary, and Lake Crescent, of beautiful, clear water, 12 miles long, with high surroundings, occupies the southeastern corner, and connects with the Saint John's through Dunn's Creek. The Ocklawaha River crosses the southern portion of the county from the west, and enters the Saint John's opposite Welaka. The portion of the county lying east of the Saint John's, and extending to Crescent Lake, is called Fruitland Peninsula, and is rich and fertile. The western portion of the county is undulating, in sections slightly hilly, with a sandy surface soil and a red and gray subsoil, interspersed with fresh-water lakes, and for cotton and general farming is the best part of the county. The pine lands will produce 10 bushels of corn or 300 pounds seed cotton, and the hammock 20 bushels of corn and 600 pounds cotton per acre, without fertilizing. Nearer the Saint John's, generally speaking, the lands are less rolling and fertile, but heavily timbered. Many portions, however, of the soil are rich in humus and other products of vegetable decomposition. The lands are generally high enough for culture. Dray-ton Island, embracing 2,000 acres, and a part of the county, is famous for its rich soil and marl. The county contains nearly every variety of Florida soil—swamp lands, high and low hammock, heavily timbered with hickory, oak, and other hard woods, and the different qualities of pine land, clay, sand, marl, and shell. A number of the finest and oldest orange groves of the State are situated in this county. There are fully 5,000 acres in the county specially devoted to the cultivation of the orange. The fruit culture and vegetable production for Northern and Western markets form a leading and profitable business, while cotton, rice, sugar, corn, and other staples are a permanent reliance for agricultural industry. There are forty-three schools, twenty-one post-offices, and more than a dozen places in the county where considerable manu-facturing and a large mercantile trade is carried on.

Palatka is the county seat, and one of the best business towns of the State, situate at the head of navigation for deep-draft steamers and sailing vessels, and near the confluence of the Ocklawaha. It possesses advantages which cannot fail of rendering it a fine commercial city. It has beautiful churches, good schools, a nunnery, and two or three of the largest hotels in the State. A narrow-gauge railroad from here to Gaines-ville, and thence to Ocala, has been completed, and is being extended to Tampa and Charlotte Harbor. Palatka is connected by telegraph with all parts of the country. At San Mateo, 6 miles south of Palatka, is an extensive orange-packing house, and the "San Mateo Institute," an excellent institution of learning, free in part.

SUMTER COUNTY.

Bounded north by Marion County, east by Orange, south by Polk, and west by Hernando, from which it is separated by the Withlacoochie

River, and has an area of over 1,300 square miles. The general characteristics of Sumter are the same as Orange, Polk, Hernando, and Marion
Counties, by which it is surrounded. The Ocklawaha River connects
the waters of Lakes Griffin, Harris, and Eustis, in the northeastern portion of the county, with the Saint John's; and Lake Pansofka, on the
west, connects with the Gulf through the Withlacoochie. A chain of
lakes in the southeastern portion of the county connects with Lake
Harris through the Pilaklikaha Creek. On the east of these lakes a
high rolling pine woods country extends for miles; on the west are fine
hammock lands and grass lands for stock. The lands in the northern
and western portion of the county also are exceptionally good, and
orange-growing is the prevailing interest. The acres planted in orange
groves are numbered by the hundreds, and the bearing trees by the
tens of thousands. No county in the State is better adapted to fruit-
growing, while stock-raising and the agricultural staples of the South
afford a sure reliance for the profitable investment of labor and capital.
Leesburg, located at the head of Lake Griffin and between that and
Lake Harris, is the county seat. The Ocklawaha River is navigable for
small steamers, and a railway extends from Lake Eustis to Astor, on the
Saint John's River.

The Tropical Railway connects with Fernandina, and the Florida
Southern with Gainesville and Palatka. Both are in process of construc-
tion south, with the view of completion to Tampa Bay and Charlotte
Harbor.

VOLUSIA COUNTY.

Bounded north by Saint John's County and the Atlantic Ocean, east
by the Atlantic, south by Brevard and Orange, and west by Orange
County, from which it is separated by the Saint John's River. It contains about 1,800 square miles. The Saint John's River runs the entire
length of its western border, and the Halifax and Hillsborough Rivers,
or lagoons, traverse the entire eastern boundary, with only a narrow
strip of land, formed by the winds and waves of the ocean, extending
between them and the ocean. It is one of the most progressive and
thriving counties in the State. The lands along the west bank of the
Halifax and Hillsborough Rivers, four or five miles wide, are the richest
hammocks, and were cultivated in sugar cane at a very early period by
the English and Spaniards, the remains of whose extensive works still
exist. It is said that four hogsheads of sugar per acre have been and
can be still produced on these lands. West of this is a belt of prairie,
interspersed with pine and cabbage palmetto, extending the entire
length of the county, and affording magnificent grazing for stock. Next,
further west, extending from the northern end of the county south about
30 miles, and varying in width from 2 to 6 or 7 miles, is a high rolling
pine country, considered by many the best for orange culture, on which
are hundreds of beautiful young groves; from this southward is a high
rolling pine scrub, until the Saint John's is reached with its varying

banks of rich hammock and savanna. Springs, lakes, and ponds abound all through the county. Enterprise, on Lake Monroe, is the county seat.

SOUTHERN FLORIDA.

South Florida, consisting of that portion of the peninsula south of latitude 28° north latitude, is composed of the counties of Hillsborough, Polk, Brevard, Monroe, Manatee, and Dade. From its low latitude, its peculiar location, as interposed between the Gulf of Mexico and the Atlantic Ocean, and its proximity to the Gulf Stream, this division has marked characteristics which specially distinguish it.

The surface is in the main flat, and, excepting the extension within its northern portion of the flattened ridge or plateau upon which the State is mostly situated, the greatest elevations found are around the external boundaries, while the depressions are in the interior, causing it to resemble the basin of a shallow lake. Thus constructed, and under the influence of the rain-bearing clouds from both sides, while the elevation of the exterior border prevents the easy egress of superabundant water, this division is not only well supplied with rivers, streams, and small lakes, but has also the broad, shallow lake of Okeechobee, and that remarkable receptacle of surplus fresh water, called the Everglades, within its borders, and occupying a large portion of its extent. It is quite probable that a clearing out of the obstructions formed in the channels of the numerous river courses reaching out from the interior to the Gulf and Ocean will relieve this section from overflow in the season of excessive rainfall, and reclaim a large territory of rich and most fertile lands, which, under the fostering influences of a climate of un-surpassed mildness, become exceedingly valuable for their immense productiveness in special crops. The savannas or grass prairies that are liable to periodical inundation during part of the year, but hidden with a rich growth of nutritious grasses during the balance of the time, form a characteristic feature of South Florida, and constitute some of the best cattle ranges in the world.

The climate is singularly equable and uniform, the difference between summer and winter being very slight, and the range of the thermometer during the year confined within very narrow limits. Warmer in winter and cooler in summer than any other portion of the State, the climate is equal to that of the most favored regions of the world, and nearly resembles that of the Sandwich Islands.

The crops in this section would not include the cereals grown with success in Northern Florida, and even corn is not grown with much suc-cess, while the apple, pear, and peach do not do as well; but, on the other hand, long-staple cotton, sugar cane, rice, jute, ramie, tobacco, in-digo, cassava, arrowroot, coffee, the olive, grape, orange, lime, lemon, citron, almond, date, cocoanut, pineapple, banana, plantain, and all the semi-tropical and tropical fruits thrive as well as in any part of the world. The keys and islands which line the coast on either side, and vary in extent from a few acres to a number of square miles, are equally available for tropical products.

BREVARD COUNTY.

Bounded north by Orange and Volusia Counties, east by the Atlantic Ocean, south by Dade, and west by Orange, Polk, and Manatee Counties. It extends along the coast for 100 miles, and contains an area over 4,000 square miles. The famous Indian River extends along its eastern boundary, the Kissimee River and Lake forming its western boundary. The climate is even and pleasant through the year. Game is plenty, and green turtle. fish, and oysters abound. Stock-raising is largely pursued; the cattle having a good range, are of good size and in good condition. Along Indian River, the west bank of which is from 10 to 20 feet above the ocean, settlements are being made. Indian River has a reputation for its oranges and pineapples, and all semi-tropical fruits, which here grow in perfection. The county seat is Titusville, a thriving town on Indian River.

DADE COUNTY.

Bounded north by Brevard County, east and south by Atlantic Ocean, and west by Monroe County. Has an area of over 5,000 square miles. Lake Okeechobee, an inland sheet of water, of over 500 square miles, without any visible outlet to ocean or gulf, occupies the northwestern corner of the county, the famous everglades the largest portion of the remainder. Along the Atlantic coast there is a strip of elevated rocky pine lands, 3 to 15 miles wide, skirted by a prairie or savanna, from a half mile to a mile in width, reaching to the everglades. This rich alluvial prairie is covered with an immense growth of grass. The climate is very equable, the extremes being from 51° to 92°. From May to October rains are frequent; during the remainder of the year there is little rainfall. In the vicinity of Biscayne the land is covered with an undergrowth of comptie, which yields an excellent article of starch and farina, similar to arrowroot. Dade County is the least populous county in the State. Miami is the county seat.

The following detailed description of this county, and the section of the State adjoining, was furnished by Lieutenant-Governor Gleason, a resident of the county, and published in a State paper. It is of interest as conveying a definite idea of the local advantages of this entire portion of the State, extending from Jupiter Inlet to Cape Sable, including the keys and islands along the reefs and everglades:

The keys are a series of islands extending along the south coast, from Cape Florida to the Dry Tortugas, lying between the mainland and the Florida Reefs, and within from 3 to 5 miles of the Gulf Stream. They are of a similar character, being of general formation and very rocky. Some are only a few acres in extent, while others contain as many as 15,000 acres. Cayo Largo is the largest. These keys are only a few feet above tide-water, and are principally covered with a growth of hard-wood timber, consisting of mastic, red and sweet bay, gumbo-limbo, crabwood, palmetto, mangrove, and a variety of oaks. The land is too rocky to admit of general cultivation, but is well adapted to the growth of cocoanuts, aloes, sisal hemp, and pineapples, all of which seem to live on a rocky soil, and grow here with but very little attention. Between these keys and the mainland is Barnes' Sound and Biscayne Bay. Barnes' Sound and Card's Sound are interspersed with innumerable small keys, covered with mangroves, and are under water at high tides, and are the resort of snipe, curlew, and other birds. In both of these sounds and Biscayne Bay are great quantities of turtle and sponges of the finest and best varieties. The sponges and turtles taken from these waters exceed $100,000 in value per annum.

The bay and all the passages between the keys and the streams running into the bay from the mainland are well supplied with a great variety of fish, such as mullet, sheepshead, grouper, &c., while incredible quantities of kingfish and Spanish mackerel are caught on the border of the Gulf Stream.

Biscayne Bay is an excellent harbor for all vessels drawing less than 10 feet of water, and can be entered at all times. The everglades are a vast shallow lake, overgrown with grass, pond-lilies, and other aquatic plants, interspersed with innumerable small islands of from 1 to 100 acres each. These islands are principally hammock lands covered over with a growth of live and water oaks and cocoa plums, with an undergrowth of morning-glories, grapes, and other vines, and are extremely fertile. The water is from 4 inches to 4 feet deep, and is very clear and pure. In many places are channels and sinks where the water is from 10 to 50 feet deep; these holes are well supplied with fish, of which the trout is the most desirable. Alligators and turtle are abundant, and panthers, wild cats, and bears are quite numerous.

Flowers of the sweetest fragrance, and of every hue and color, greet the eye. The border and outer margin of the everglades is prairie of from one-fourth to one mile in breadth, and comprises some of the finest and richest land in America, having once been a portion of the everglades, and formed by the receding of the waters. The soil is sandy, with a mixture of lime and vegetable matter, and freely effervesces when brought in contact with acids.

The strip of land between Biscayne Bay and the everglades is from 3 to 15 miles in breadth, and is principally rocky pine land, with an undergrowth of a species of sago-palm, called by the Indians "koonitie," which name has been generally adopted by the whites. It makes a very good article of starch, and excellent gavini, which cannot be distinguished from Bermuda arrowroot except by microscopic tests.

This section of the country has evidently been an uplift or upheaval, as the rock dips at an angle of about twenty-three degrees, and slopes both toward the bay and the everglades. The rock, in many places, is in circular form, and is coral. The soil is sandy, which, mixing with the decomposed lime of the coral rock, forms an excellent and inexhaustible soil for grapes and sugar cane. The country north of Biscayne Bay, towards Jupiter Inlet, is of a similar character to that already described, with the exception that there is no rock. Fine springs of water are found in different localities, and burst forth with great force; some of these are mineral springs, principally chalybeate. Sea-island cotton is grown here, and it is a perennial, and can be picked several times each year. Grapes flourish well, and are not subject to mildew, and ripen about the middle of May. Tobacco raised along the bay will compare with the best of Cuba. Bananas, plantains, oranges, coffee, dates, pineapples, rice, indigo, sugar, apples, arrowroot, cassava, all grow and thrive, as well as the garden vegetables of the Northern and Middle States. Indigo, when once sown, remains in the ground and ratoons as it is cut off. Sugar cane ratoons, and requires planting only once from four to five years. Sugar cane can be raised here with less labor than in Cuba, as the land is easier cultivated, and a sugar plantation can be made for one-fifth of the money which it can in Louisiana.

This section of the State is capable of producing all the products of the West Indies, and there is no doubt that when this portion of the country becomes known it will be rapidly developed. Sea-island cotton can be raised with half the labor that is required in the northern part of this State or in South Carolina, as this is beyond the region of frost. The climate is very agreeable, being tempered by the Gulf Stream. It is not as warm here in summer as in New York, or as cold in winter as in Cuba, as there are no mountains or high elevations of land. The thermometer averages 73°, and the extremes are 51° and 92°.

There is a constant sea-breeze off the Gulf Stream, commencing about 8 o'clock a. m. and lasting until nearly sundown. The climate is very exhilarating, and a white man can do as much labor in a day as in any portion of the United States.

The constant Indian wars, which have been more severly felt in this county than in any other portion of this State, have retarded its growth and prevented its development. Biscayne Bay is within four days of New York, and is the best locality in the United States for raising vegetables and fruit for that market. All kinds of vegetables can be raised in the winter, and pineapples and limes are three weeks earlier at this place than in the Bahamas or Cuba.

HILLSBOROUGH COUNTY.

Bounded north by Hernando, east by Polk, south by Manatee County, and west by Gulf of Mexico. It contains about 1,500 square miles, embracing Clearwater Harbor, Tampa and Hillsborough Bays, with the Hillsborough, Alafia, and Little Manatee Rivers entering from the north and west, and many keys or islands on the coast. The land lies more level than in Hernando County, and though generally lighter, is still fertile. Tampa City, a port of entry situated at the head of Tampa Bay and mouth of Hillsborough River, is a thriving place, and the county seat. The citizens are enterprising, and the cultivated lands and orange groves in city and vicinity show constant progress. Cattle-raising for export has always been a leading business in this and adjoining counties. Large numbers are exported annually to Cuba. Recent immigration and increased population has led to very extensive cultivation of oranges and semi-tropical fruits and vegetables, for which there is no better section, as climate, soil, and transportation are favorable. Some tropical fruits have been successfully cultivated. The usual Southern staples of cotton, cane, and rice are raised, as also field crops of all varieties common to other sections. There are now five lines of railroads in contemplation, three of which are in process of rapid construction, from the Saint John's River and the Atlantic ports, with Tampa as the objective point.

MANATEE COUNTY.

Bounded north by Hillsborough and Polk; east by Brevard and Dade Counties, being separated from the latter by Lake Okeechobee; south by Monroe County, and west by the Gulf of Mexico; containing an area of over 5,000 square miles, and embracing the northern portion of Charlotte Harbor, the southern portion of Tampa Bay, Sarasota Bay, and the numerous islands adjacent. Peace Creek, rising in Polk County, subdivides the county near the center, and runs south to Charlotte Harbor, having numerous tributaries, which, with many lakes, water the interior of the county. The Myakka River discharges into the harbor further westward, and the Manatee River, in the northwestern part of the county, enters Tampa Bay. The surface is generally level, lands light, piney woods, hammocks, and prairie. Pine Level is the county seat.

Along the rivers and borders of lakes the land is very productive; a large portion of the country is given up to stock-raising, which is a leading and profitable business. Over 100,000 head subsist at no cost or care, except the gathering to brand and mark, or for sale and delivery. Key West, Cuba, and other is lands afford a constant and good market, and steamers and vessels are regularly engaged in the transportation, mostly from Tampa Bay and Charlotte Harbor. There are many stockmen who count their herds by the tens of thousands. Peace Creek, a large stream, is susceptible of steam navigation through the county, and is attracting immigrants, especially those who seek an equable climate, and to locate below what is called the frost-line. Long-staple cotton, cane, rice, tobacco, do well, and will become leading staples of export.

MONROE COUNTY.

Bounded north by Manatee, east by Dade, south and west by the Gulf of Mexico. It has an area of about 5,000 square miles, a large portion of which is occupied by what is known as the Big Cypress Swamp and the Everglades. The county includes the numerous keys and islands off the Florida southern coast, the most important of which is Key West, where the principal wealth and population are located, and the southern portion of Charlotte Harbor, Pine, Sanibel, and other islands. The Caloosahatchie traverses the northern portion of the county and enters Charlotte Harbor, and is navigable as high up as Fort Thompson, some 30 miles from its mouth. It is contemplated to connect this river with the great lake Okeechobee and drain the extensive country about the lake. The northern portion of the county is adapted to semi-tropical and tropical fruits, and also many of the keys, which are already famous for cocoanuts, pineapples, and bananas. Key West is the county seat.

POLK COUNTY.

Bounded north by Sumter and Orange Counties, east by Orange and Brevard, south by Manatee, and west by Hillsborough. The Kissimmee Lake and River separates it from Brevard. It has an area of about 1,900 square miles, and its general characteristics are the same as Sumter and Manatee. The surface is undulating, the lands hammock, pine, and prairie, dotted all through with small lakes of clear water abounding in fish. The prairies are the range for herds of cattle, of which there are 100,000 head in the county. Bartow, situated on Peace Creek or River, is the county seat, and a thriving business place. The lands within 2 or 3 miles of Peace Creek and its tributaries are excellent farming lands and well adapted to the culture of the orange and semi-tropical and some tropical fruits.

PRODUCTIONS OF FLORIDA.

Having thus presented a general survey of the climate, soil, and topography of the State, in which I have made free use of the material furnished through official publications and well-authenticated sources, I next proceed to consider the productions and agricultural capabilities of Florida. In this I shall avail myself of the practical experience of those who have labored to develope these resources.

So much has been said of the vast scope of vegetable growth in Florida that it is a cause of surprise to all strangers and suspicion to many, as though there might be well-grounded suspicion of exaggeration or overstatement. On this account it is deemed best to refer to some of the older standard writers on this subject.

In the "Observations" of Charles Vignolles, published in New York in 1823, on page 99, we find the following:

The following list of productions capable of being raised in Florida has been made out with some pains, and it is believed all these stated are profitable and practicable articles:

Oranges, various kinds,	Currants, Zaute,	Cinnamon,
Lemons,	Pineapple,	Pimento,
Lime,	Figs,	Sago palm,
Citron,	Plantain, ·	Red pepper,
Shaddock,	Banana,	Saponica,
Mango,	Yam, ·	Jesuit's bark,
Pawpaw,	Bread-fruit,	Besine,
Cocoa,	Arrowroot,	Palma Christi, castor-bean,
Dates,	Gallnuts,	Tea,
Sweet Almond,	Doliahos, or soy-lean,	Sugar,
Bitter Almond,	Jalap,	Tobacco,
Pistachio,	Tree rhubarb,	Rice,
Acuaqua,	Ginger, ·	Cotton,
Gum gleni,	Gum guiacum,	Silk,
Fustic,	Braziletto,	Cork-oak,
Balsam,	Senna,	Chestnut,
Hemp,	Turkey madder	Sassafras,
Camphor,	Balm of Gilead,	Sarsaparilla,
Frankincense,	Cloves, ·	True opium poppy,
Leeche plant of China,	Liquid-amber,	Tumeric,
The olive,	Aloe,	Nutmegs.
The vine, all varieties,		

Such a list seems wonderful enough as attributed to a single State, but this list, made fifty years ago, is far from comprising all the products which may be profitably cultivated, or are within the capacity of this soil and climate.

Besides the staples common to more northern latitudes, cotton, wheat, rye, oats, peanuts, cow-pease, Irish and sweet potatoes, melons, and all the variety of garden vegetables, and the strawberry, blackberry, huckleberry, plum, pomegranate, and quince, we may add coffee, cassava, indigo, cochineal, Sisal hemp, the guava, tamarind, sapadillo, avocada pear, mamie-apple, custard-apple, pecan-nut, &c.

The year following the acquisition of the territory of Florida, 1822, a French gentleman, Peter Stephen Chazotte, presented to Congress a memorial setting forth the advantages of the climate and soil for tropical productions, and asking that the government allot one thousand acres of land in the southern portion of Florida, with an appropriation of $50,000, for the establishment of an experimental farm and the introduction and propagation of coffee, cocoa, and other products of tropical countries.

At a still earlier period, a proclamation of George III, issued from the Court of St. James, 7th day of October, 1763, and by the authority of a treaty of peace concluded at Paris on the 10th day of February, the same year, assigned to Colonel Grant—

The government of East Florida, bounded to the westward, by the Gulf of Mexico and the Apalachicola River; to the northward by a line drawn from that part of said river where the Chattahoochee and Flint Rivers meet, to the source of the Saint Mary's River, and by the course of said river to the Atlantic Ocean; and to the eastward and southward by the Atlantic Ocean and the Gulf of Florida, including all islands within six leagues from the sea-coast, with the expectation that rice, indigo, silk, wine, oil, and other valuable commodities would be produced in great abundance.

The English at that time knew little or nothing about coffee, as its cultivation at that period was confined to St. Domingo, and had not been introduced upon the Island of Jamaica. At a later period, an English gentleman of fortune went to establish himself in East Florida, and entered successfully into the culture of coffee and sugar-cane, and his establishments were already considerable when the American Revolution, in its effects, caused Florida to pass into the hands of Spain. The British Government, finding he had so far succeeded, would not allow him to remain, but destroyed his plantation, and carried him off with his slaves, awarding him a considerable sum for his loss and damages.

Mr. Carvert says:

So mild is the winter that the most delicate vegetables and plants of the Carribee Islands experience there not the least injury from that season; the orange tree, the plantains, the guava, the pineapple, &c., grow luxuriously. Fogs are unknown there, and no country can, therefore, be more salubrious.

Mr. William Stork, in his description of East Florida, gives the following account of it:

The productions of the northern and southern latitudes grow and blossom by the side of each other, and there is scarcely another climate in the world that can vie with this in displaying such an agreeable and luxuriant mixture of trees, plants, shrubs, and flowers. The red and white pine and the evergreen oak marry their boughs with the chestnut and mahogany trees, the walnut with the cherry, the maple with the campeach, and the braziletto with the sassafras tree, which together cover here a variegated and rich soil. * * * The wax myrtle tree grows everywhere here. * * * Oranges are large, more aromatic and succulent than in Portugal. Plums naturally grow finer and of a quality superior to those gathered in the orchards in Spain. The wild vines serpentine on the ground, or climb up to the tops of trees. Indigo and cochineal were advantageously cultivated there, and in the year 1772 produced a revenue of $200,000.

Chazotte adds, in 1822 :

This country will produce all the tropical fruits and staples by the side of those belonging to a northern climate.

This fact is now being demonstrated in the experience of the recent population of the State.

In impressing upon the attention of Congress the advantages of entering upon the cultivation of coffee and cocoa, or the chocolate plant, in connection with vines, olives, capers, and almonds, Chazotte gives the following statement of feasibility and profits of cultivation :

Coffee.—One acre of land planted by ranges, and the plants at 5 feet distant from each other, gives 1,764 plants. A man can take care of two acres, which gives 3,528 plants. Each plant may, by an average, yield 2 pounds or more, but I reduce it to 1 pound ; therefore, a man will give yearly 3,528 pounds of coffee, which, at 25 cents, produces $882.

It is to be observed that no crop is to be expected on the first and second years ; on the third year the plant yields a good crop ; on the fourth an abundant one, which it will continue to yield every year until the ground is exhausted and the plant dies. For the two first years of the planting, all kinds of vegetables and corn may be planted between the ranges ; they will yield two crops in one year. Cotton is not to be planted between the ranges.

Cocoa.—Four acres of land planted in rows, and the trees at 10 feet distant from each other, give 1,764 trees. A man is capable of taking care of them and of gathering the nut. At seven years of age each tree will yield two pounds, and the quantity will increase with its age ; therefore, a man will gather 3,528 pounds of cocoa, which, at 15 cents per pound, will produce $529.20.

This cultivation, differing from all others, requires some illustration. It was formerly thought that its culture required much labor and a virgin soil ; but experience has shown that it grows on land half exhausted by the coffee-plant, and in less than twelve years' time acquires such power as to destroy the coffee underneath. Hence it is now planted between the ranges of coffee when this last is about seven years of age ; so that when the land would otherwise become a mere waste, requiring a hundred years for forests to rise on it again ere it could recover its first fruitfulness, the same land being again covered by a new forest of productive trees, the fruits of which growing and maturing all the year round, each day brings in its crop.

The extraordinary effects of the cocoa tree in regenerating the ground upon which it grows may easily be accounted for. This tree seldom attains higher than fifteen feet ; it is branchy, its leaves very large, and the body, or stock, of a middling size ; the leaves continually falling off the tree while new ones grow, cover the earth with a thick bed of leaves, which allow not even a blade of grass to grow with them. Hence the ground requires no culture, and the trees but a light pruning when any ravages have been caused by storm. This constant thick bed of leaves returns to the earth five times more nutriment than the diminutive size of the tree requires from it, and in less than thirty years brings the soil back to its original fertile state.

Having given the proceeds of a man's yearly labor in the plantation of coffee and cocoa, I shall now quit Florida and enter the territory of the United States.

Vines.—An acre of land planted with vines, allowing 41 ranges at 5 feet distant, and to each range 104 vines at 2 feet apart, gives 4,264 vines to an acre. Five acres for a man's labor give 21,320 vines ; and allowing the grapes of 10 vines to yield 1 gallon of

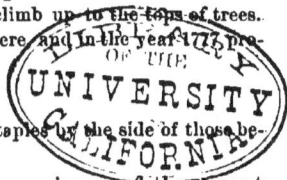

wine, it will produce 2,132 gallons, which, being rated at the low price of 30 cents per gallon, will produce $639.30 for a man's yearly labor. As olives, capers, and almond trees require no particular culture, they may be planted in ranges, at 30 feet distant, in the vineyard, where the mildness of the climate allows the plantation. In Georgia and Alabama these four productions may be raised on the same soil.

Vines, olives, capers, and almonds, planted on the same ground:

5 acres in vines produce, as before stated, for a man's yearly labor	$639 30
175 olive trees, at thirty feet distant, will yield, after seven years of age, about one gallon of oil each, which, valued at the low price of $1.50 per gallon, is	262 50
45 almond trees, 25 caper trees produce, valued at $1.50	105 00
Yearly proceeds of a man's labor	1,006 80

I shall now suppose that, in the course of 30 years, we may employ 50,000 persons in the culture of vines singly:

They will cultivate 250,000 acres of land, which will yield an annual revenue of..	$31,965,000
50,000 persons in the culture of vines, olives, almonds, and capers, on 250,000 acres of land, will yield an annual revenue of	50,340,000
Total	82,305,000

Cocoa.—Fifty thousand persons engaged in the culture of cocoa will cultivate 200,000 acres of land, which will yield an annual revenue of $26,420,000.

Coffee.—One hundred thousand persons engaged in the culture of coffee will cultivate 200,000 acres of land, which will yield an annual revenue of $88,200,000.

Recapitulation of the preceding estimates:

100,000 persons cultivating 500,000 acres in vines, olives, &c., produce..	$32,305,000
50,000 persons cultivating 200,000 acres of cocoa........................	26,420,000
100,000 persons cultivating 200,000 acres of coffee......................	88,200,000
250,000 persons cultivating 900,000 acres, produce	196,925,000

The home consumption of this country may be estimated to be annually:

Wines, olives, &c	$17,305,000
Cocoa, &c	6,420,000
Coffee	13,200,000
Home consumption	36,925,000

Leaving an immense surplus of exportation to foreign countries of—

Wines, olives, &c...	$65,000,000
Cocoa	20,000,000
Coffee	75,000,000
Exportation	160,000,000

FIELD CROPS—SUGAR CANE.

Florida, both in climate and soil, is peculiarly well adapted for growth of cane ; the earliest colonists cultivated it, and successive occupants, French, English, Spanish, American, have grown it successfully; the long period of warm weather, and the absence of cold, give a longer period for the cane to mature. During the English occupation many large plantations were opened, and later, since Florida became United

States territory, there have been several large sugar plantations profitably carried on.

Latterly, cane has only been planted for domestic use and neighborhood sale. But, even rudely raised and rudely manufactured, Florida sugar and sirup rival, in color, grain, and quality, the best Louisiana.

Fair land will produce from 1,500 to 2,000 pounds of sugar; rich land, thoroughly fertilized, will produce from 2,000 to 4,000 pounds. Recent improvements in sugar machinery have obviated the necessity of expensive works formerly required, rendering it possible for the small as well as large planter to manufacture cheaply; its cultivation is as easy as corn, and its immunity from all hurt by ordinary enemies to other vegetation renders it a safe crop.

The superiority of Florida over any other section of the United States in adaptability to the growth of cane is mainly based upon her milder climate, the greater length of the seasons, and the correspondingly longer growth and larger size of the cane. In Louisiana from three to five feet may be taken as the average size of cane when harvested, while in Florida from five to seven may be taken as the average size of the cane over the whole State, extending from north to south nearly 400 miles, and, with fair culture, eight, ten, and twelve feet are quite common lengths. As early as 1823, Vignolles wrote:

Respecting sugar, the recent successful trials that have been made upon it have determined the curious fact that it will grow in almost any of the soils in Florida south of the mouth of the Saint John's River; the great length of summer, or period of absolute elevation of the thermometer above the freezing point, allows the cane to ripen much higher than in Louisiana.

Williams, writing in 1837, says:

This (sugar) ought to be the staple of the country. Experiments in every part of the territory prove that all our good lands will produce sugar cane as well as any other crop. * * * A general impression has prevailed that sugar could not be made to advantage unless a great capital is invested; but experience abundantly proves that a small capital may be as profitably employed in the culture of cane as in any other product.

In an article on sugar cane in the new American Encyclopedia the climatic disadvantages attending the cultivation of the sugar cane in Louisiana are stated as follows:

Yet, the climate of Louisiana itself is rather north of that best suited to the plant, the cane being frequently killed by the frost after starting in the spring, and at maturity in the latter part of October and in November, the effect of which is to materially diminish its production of sugar. In 1857, injurious frosts thus occurred in April as late as the 22d, and on the 19th and 20th of November. In November, 1859, the cold was very severe on the 12th, 13th, 14th, and 15th, in all parts of Louisiana, the thermometer on the 14th standing at 25° F. at New Orleans, and thick ice being formed in the most southern parishes. The effect of this was that the cane was everywhere frozen, and land which had previously given above two hogsheads to the acre yielded barely half a hogshead, and this of inferior quality. The climate is also subject to long-continued drought, which seriously injures the growing crops.

But in Florida frosts are of infrequent occurrence, and in South Florida are unknown. Of the few frosts that do occur, instances as early as

November or as late as April have been known only at intervals of years.

With the protection against competition with foreign cheap labor now afforded by the government, sugar will speedily become one of the commanding industries of Southern Florida especially, and a source of immense wealth to the State.

Dr. Westcott, president of the Madison County Agricultural Society, in 1870, gave the following on the subject of cane culture:

It takes about the same labor to cultivate a sugar-cane crop as it does for corn. For a farmer not cultivating more than 5 or 10 acres of cane, the expense of an iron mill, boilers and brickwork, house or shed, &c., would not cost to exceed $400. To manufacture 10 acres of cane would require the work of six men forty days; one pair of mules, horses, or oxen at the mill, and another pair to haul the cane from the field. The profits of 10 acres planted in cane, from actual experiment, omitting capital required for boilers, mill, troughs for crystallizing, houses for draining, teams, &c., are as follows:

DR.

Ten days' work of team to break up land, at $1.50 per day	$15 00
24,000 seed cane, at $10 per M	240 00
Fifteen days' work planting, at $1	15 00
Ten days' work with hoe	10 00
Fifteen days' work with cultivators and plows	22 50
Six men 40 days, equal to 240 days' work, manufacturing, at $1	240 00
Two pair oxen 40 days, at $3 per day	120 00
Barrels, &c	60 50
	723 00

CR.

By 3,700 pounds sugar per acre, 37,000 pounds, at 10 cents	3,700 00
Showing a net profit of	2,977 00

It is no uncommon thing to produce, by proper fertilizing, 2,000 pounds of sugar and 170 or 200 gallons of sirup, equal to 1,700 pounds of sugar, or a total of 3,700 pounds of sugar, of a superior quality, per acre.

Sugar requires natural strong land, or well-manured light land, the latter making a better quality of sugar. By properly manuring the ratoon, or cane springing up from the root, after the first crop from planting, it will yield nearly the same product for two or three years; after that time experience teaches it is best to remove the roots to other ground. It will be observed that after the first planting there is no more expense for seed cane.

COTTON.

Sea-island or long cotton is raised mostly from the Suwannee River to the ocean, and south of latitude 30°. The average product per acre is from 150 to 200 pounds, though it often exceeds double that. This species of cotton is only raised on the sea islands bordering South Carolina, Georgia, and in Florida, our State raising over half the total crop. The price ranges from 25 to 50 cents per pound, though there are planters who readily get more than these figures, but their cotton is exceptionally fine. Short cotton is grown west of the Suwannee to the western and northern boundaries of the State. It will average from 200 to 500 pounds to the acre. In grade, Florida cotton rates with the best. Cotton raising, however, is subject to some risks; cold, rain, drought, or caterpillar often sweeps localities. Generally speaking, it is a safer crop in Florida than anywhere else. New methods of cultivation, improved seed, remedy for the caterpillar, are adopted by the intelligent and prudent planter, who is not subject to a loss which a careless, shiftless man may have. The methods of cultivation are simple, the crop itself affording by its seed the very best fertilizer. As the seed is fully 75 to 80 per cent. of the cotton as picked, it is largely sold and exported, and its increasing value for manufacture renders the cotton crop profitable, even at the present low price of the staple. At the late cotton exposition in Atlanta, Ga., a bale of long cotton from Florida was pronounced by the foreman of the Willimantic Thread Factory as the best in the entire collection for his purposes, taking the first premium.

RICE.

Rice, which constitutes the main food of the great majority of the population of the world, is raised here mostly for domestic use. There are thousands of acres in every section of the State peculiarly adapted to its successful culture. Its cultivation is as simple as any cereal; usually drilled and kept clear of weeds, 25 to 75 bushels of rough rice is a fair crop. Recent introduction of improved rice machinery, adapted for individual and neighborhood use, will stimulate increased production. Limited by climate, rice will always prove a remunerative crop. It is generally supposed that rice is only successfully grown on low lands which adjoin tide-water, and can be overflowed at certain different stages of growth. It is true that the great bulk of the crop is grown in this way, but more recent experience has demonstrated that it can be grown successfully upon dry land, and upland rice is now becoming one of the reliable and profitable field crops, more remunerative even than wheat at the North. The upland rice from this State displayed at Atlanta received the first premium.

A low, moist soil has generally been planted; overflowing is not needed, but on any good land it is successfully cultivated. It has needed only introduction of rice-cleaning machinery to make its cultivation universal in Florida. Quite recently a company of practical business men have been formed, who have put up extensive works, which will be able to receive and prepare all that may be raised. It may be relied upon as one of the permanent staple products of Florida, which will add largely to the exports of the State and afford a sure and profitable reliance for the farmer.

CORN.

Corn, which is the great food staple raised in the United States, especially in the West, and which exceeds by many millions of bushels any and all other crops, is grown in all portions of the State, and the produce per acre is here, as elsewhere, more or less, according to fertility of soil and cultivation. Ordinary pine land will produce, say, 10 bushels ; good hammock land, 20 to 50 bushels, according to the cultivation. Extra culture here, as everywhere, will largely increase the product. Ex-Governor Drew has raised, near his mills at Ellaville, in Madison County, 120 bushels of superior corn to the acre. Corn here is planted in February to April, plowed at intervals, laid by in June and July ; blades stripped for fodder, and stalks with ears left in field to be harvested at leisure. It may be cribbed in field in the shuck, suffering no damage from weather, or housed in corn-crib near the dwelling; shucked and shelled if for sale or food. When fed to stock it is fed in shuck. One person with one mule can easily cultivate from thirty to forty acres, and as the time for planting to final plowing is only from four to five months, it leaves ample time to cultivate another crop of pease or sweet potatoes with same labor on same land. The corn usually raised is the white variety, largely used in meal and hominy for food, especially at the South. The Northern farmer who has been used to see 40 to 60 bushels ordinarily raised on the old homestead, should, in comparing the relative production South and North, take into consideration cheapness of land, number of acres which can be cultivated, time taken to make crop, expense of gathering, saving, housing, and also value, transportation, and its quality. White is best for food. All things considered, corn is one of the most useful and profitable crops to raise in Florida.

WHEAT, RYE, OATS.

Wheat in the northern section of the State is grown to some extent, but is not generally raised as a regular crop. Rye and oats do well, and are mostly sown early in the fall, affording a good winter pasturage ; mature in early spring, and are not thrashed, being cured and fed to stock in the straw.

PEANUTS.

The peanut, pinder, goober, or ground pea, as the plant is variously called, grows well on almost any warm, light soil. The seed should be planted early in the spring. The after cultivation is simple. A hundred bushels to the acre is an average crop. They are worth $1 to $2 per bushel. The nut produces an oil which is said to be equal to the finest olive oil.

PEAS.

The common English pea is not cultivated as a field crop, but as a garden product is largely grown for the winter market, and affords great profits. The cow pea is extensively grown and produces excellent crops.

It resembles the bean family in the appearance of its foliage and the manner of its growth. It is common to sow them between the rows of corn at the last plowing. They will produce from 10 to 15 bushels per acre, besides a large amount of forage. On account of the luxuriant growth of the vine, on poor soils even, its culture as a green crop, to be turned in, is fully as advantageous as clover at the North.

TOBACCO.

Tobacco will grow anywhere in the State. A superior quality of Cuba tobacco, from imported seed, is mostly grown in Gadsden and adjoining counties, and fully equals the best imported. Before the war it was extensively and profitably cultivated, and mostly sold to Germany, agents visiting the State to purchase. It requires careful attentin, will yield from 500 to 700 pounds to the acre, and sells for from 20 to 30 cents per pound. Latterly there is an increasing home and State demand by cigar manufacturers, and the area of cultivation is extending.

SWEET POTATOES.

This crop, as an article of food, is as universal in all Southern households as rice is to the Chinese, macaroni to the Italian, or the Irish potato to the Irishman. White or black, no family is so poor but what has a potato patch. It yields all the way from 100 to 400 bushels to the acre, according to soil, cultivation and season; is grown from roots, draws, and slips; planted from April to August, and maturing from July to November; is of easy cultivation, and may be dug and safely banked in field and yard, or housed; is eaten raw or cooked, and the old-time cook can make most appetizing dishes of it. There are many varieties planted, good and indifferent, and there is no excuse for not raising the best. It may be raised at a cost of fifteen cents a bushel, and brings in the home market from 40 cents to $1.

IRISH POTATOES.

This crop does not produce as well as at the North, but is off in time to be followed by a crop of sweet potatoes the same year. They should be planted in December or January, although good crops are sometimes obtained from later planting. A covering of muck, grass, or coarse compost is very beneficial. The potatoes are fit for digging in May. They can be shipped without difficulty, and at a moderate expense, to the Northern markets, where they are worth from $5 to $9 per barrel. The culture is essentially the same as that practiced at the North.

ARROWROOT, CASSAVA, COMPTIE.

All these grow well when cultivated, and produce astonishingly. Florida arrowroot grades in quality and price with the best Bermuda. Cassava, from which starch and tapioca are made, attains great size.

3290——4

Comptie, the bread root of the Indians, grows without any cultivation. All of the above have only been grown for domestic use for starch and for food, and have limited sale in this and adjoining States. The attention of Northern starch manufacturers has lately been drawn to them, and Governor Sinclair, of New Hampshire, having tested the roots by actual experiments, has introduced a pioneer factory in Orange County. As either and all of these roots have a larger percentage of starch in them than the Irish potato, and can be grown at the same price, and manufactured all the year, we may look for a large business in this industry.

<h3 style="text-align:center">SISAL HEMP, RAMIE, JUTE.</h3>

All of the fibrous plants grown in warm latitudes are found here. Some years ago the Sisal hemp was largely grown, but the Indian war broke up the country where it was planted, and the cultivation has not been resumed to any extent. A Key West writer says that a ton of fiber may be grown to an acre, worth $300. Extensive preparations are being made for the cultivation of jute in South Florida, and at no distant day it will become a leading industry. Col. A. B. Lindermann, of Philadelphia, is at the head of a company, recently organized, to test the cultivation of jute and indigo upon a large scale. An ample supply of seed has been imported from India; suitable lands have been selected from the Disston purchase, in Sumter County, and arrangements are now being made to procure the necessary workmen. An expert will visit the crop during the coming summer, and should his report be favorable, a large amount of English capital will at once be invested in the business.

A variety of wild jute is found growing abundantly in East and South Florida. A sample of the fiber of this plant, roughly prepared, was taken to Dundee, Scotland, where the principal jute factories of the world are located, and was valued by the proprietors of the works at $90 per ton. Two crops can be readily grown during the season, and improved machinery has lately been devised for the preparation of the fiber. The plant grows vigorously upon low, wet soils, is difficult to eradicate when once planted, and promises to add another to the many flourishing industries of Florida. It is believed by those who have investigated the subject, that this State is capable of furnishing all the jute required for consumption in this country, now imported from Calcutta at a cost of many millions annually.

<h3 style="text-align:center">INDIGO, CASTOR BEAN, AND SILK.</h3>

The indigo plant is indigenous in Florida; during the English occupation it was extensively cultivated, manufactured, and exported; now it is occasionally made for domestic use. The castor bean here attains the size of a tree often 30 feet high, grows rapidly, and bears largely; now only used for home purposes. Silk some years ago attracted a good deal of attention, but is now only occasionally produced as a pastime.

The different species of mulberry grow here to perfection from root, cutting, or graft; in leaf from March to October. In time, no doubt, the business will become a regular industry. The company above referred to intends to engage largely in the culture of that valuable dye, indigo. About the year 1770 this article formed the principal export from Florida, and the old works still to be seen in the vicinity of New Smyrna, on Indian River, indicate the vast extent of the plantations devoted to this enterprise.

FRUITS.

The most promising and fascinating industry, now absorbing attention, more particularly in East and South Florida, is the cultivation of fruits, of which the citrus family takes the first rank. This group comprises all the varieties of the orange, citron, lemon, lime, and shaddock, numbering more than a hundred.

Dr. Sickler, who spent six years in Italy, and paid great attention to the kinds and culture of the citrus, published at Weimar, in 1815, a quarto volume, called Volkommene Orangerie Gartner, in which he describes seventy-four sorts. He arranges the whole into two classes, and these classes into divisions and subdivisions, without regard to their botanical distinctions or species, as follows:

Lemons:	Sorts.
Cedrats, or citrons	4
Round lemons	6
Pear-shaped lemons	11
Cylindrical lemons	4
Gourd-shaped lemons	2
Wax lemons	5
Lumies lemons	8
Cedrat, lemons or citronate	6
Limes	4
Oranges:	
Bitter oranges	6
Sour oranges	6
Sweet oranges	12

Few other classes of fruits are more easily propagated than the citrus, and all of the species may be rapidly increased and produced either by seeds, cuttings, layers, grafting, or budding, the lime being the most difficult and the citron the most easy of propagation. They differ from deciduous fruits in the respect that like always produces like, the seed of every variety invariably producing its kind. Cuttings of thrifty wood, two years old, strike fibers as rapidly as younger wood, though the mode of propagating almost universally adopted in Florida is by budding upon young stocks from the nursery, or from the larger stocks obtained from the forests. The citrus family of fruits is supposed to have originated in the warmer parts of Asia, and to have derived its name from the town of Citron, in Judea, though it has been cultivated from time immemorial in middle and southern Europe, and is now cultivated almost throughout the world, and in no higher degree of perfection than in East Florida, south of the 30th degree of north latitude.

THE ORANGE.

The cultivation of the orange (*Citrus aurantium*) in East Florida, previous to 1835, had attained a degree of considerable commercial importance, and the exports of this fruit from the small city of Saint Augustine are said to have amounted to $100,000 annually. On the Saint John's River, and in some parts of West Florida, and at Tampa Bay, groves were being established as a source of commercial supply. In February, 1835, a very severe frost visited the State, and most of the orange groves and other semi-tropical fruits were destroyed, or nearly so, leaving only the stumps and roots to spring up again. Many of these sent up shoots, and began to encourage hopes of returning prosperity to this branch of industry. These hopes were not permitted to be realized, however, for in 1842 an insect called the *Orange coccus*, or scale insect, appeared in the orange groves, and spread with great rapidity over the whole country, almost totally destroying every tree attacked.

This calamity continued for ten or twelve years, and bade defiance to almost every effort made to stay its blighting force. Many became discouraged in the contest and abandoned further attempts to re-establish this heretofore agreeable and profitable branch of industry. In 1853, however, the insect began to decrease in numbers, and finally disappeared, since which time most of the groves now in the State must date their birth. These groves are rapidly multiplying in all parts of the State, and the bearing trees are now numbered by tens of thousands, while the young groves, which are being constantly started, comprise millions of trees. The orange from the seed produces fruit in from seven to ten years, depending upon situation, culture, &c.

Groves made from wild stocks, usually cut off at a height of 3 to 4 feet from the ground, and the new shoots budded, generally produce fruit in three years. The number of oranges produced from a single tree varies from 100 to 10,000, according to the age, situation, and treatment of the tree. The trees are usually set 20 feet apart, and an acre will contain about 100 trees. Florida oranges were usually sold, previous to 1835, at $7.50 to $10 per thousand. Now a demand exists for twenty times our present supply, at $15 to $20 per thousand, as they hang upon the trees.

HOW TO MAKE AN ORANGE GROVE.

The judicious selection of the land is the first and most important point, for on this success in a great measure depends. Choose high, dry hammock or high rolling pine land that has natural drainage and a yellowish subsoil. The low, flat lands which are underlaid with hard-pan or sandstone, mixed with oxide of iron, require ditching or draining and much care in setting the trees, so that the roots may have free scope and relief from standing water. The most favorable locations are on southeast side of wide sheets of water, or high lands, which are more generally free f.om frost. The land selected, clear thoroughly of all trees, &c., break up well, and substantially fence; sow with cow peas, which turn under when in bloom—it improves and sweetens the soil; this may be done before or after planting trees. Dig holes 30 feet apart, 18 inches deep, and 4 feet in diameter; clean out all roots; fill up with top soil, which will retain the moisture; procure trees from three to five years old, take them up carefully, with all of the roots possible, pack up with wet moss as soon as dug, put in shade and out of the wind, take to the proposed grove carefully; remove soil from holes dug sufficient for the tree, with roots carefully spread, trunk standing in same position as originally grown. Let the tree, when set out, be fully an inch above natural level of land; fill under, in and about the roots, compactly—it is best done by the hand, filled to surface and gently tramped down; fill on some 2 or 3 inches of earth, which will prevent drying; the rainy season commencing, remove the soil about the tree to the level about it. Cultivation should be frequent and shallow, and trash not allowed to accumulate near trunk; light plowing and raking near the trees is best and safest. Following these general directions, no one should fail. The cost of a five-acre grove, at, say, five years from planting, at a liberal estimate where high pine land is chosen, will be about as given below. If hammock land is taken, the cost of clearing will be more. The grove will have begun to yield at the end of the period named. Rev. T. E. Moore, Fruit Cove, Fla., has published a good treatise on orange culture.

COST OF GROVE.

Five acres of good land, variously estimated, depending on location.

Cutting timber, clearing	$75 00
Fencing (post and board fence) and breaking up	75 00
Three hundred trees and setting out	200 00
Manures, labor, cultivating, taxes, &c., for five years	500 00
Total, less cost of land	850 00

Such a grove would readily sell now in Florida for $1,000 per acre. From and after five years the annual growth of trees and increase of fruit is constant for at least ten years, and the grove will hold its vigor and fruit-producing qualities for a century or more. The orange is a hardy tree, will stand great extremes of rain and drought; it will show the effects of a single season's neglect, and quickly show a single season of care and attention.

THE LEMON.

The lemon is produced in East Florida to a degree of perfection far surpassing the same fruit grown in the West Indies, Sicily, Italy, or

Spain, and persons familiar with this fruit in those countries are rather disposed to discredit the statement that the lemons of Florida are of the same variety of fruit. The Sicily lemon grown here frequently reaches from 1 to 2 pounds in weight, and is of a superior quality.

VARIOUS FRUITS.

There are five of the acid varieties of the lime (*Citrus limetta*) named in English nursery catalogues. The juice of the lime is preferred to that of the lemon as being more wholesome and agreeable, and when freely used is a preventive of fever. Combined with a little salt, it is regarded by some of the old settlers as a specific for chills and fever.

The citron (*Citrus medica*) is commercially known in the United States as a preserved confection, imported from the Mediterranean in oblong boxes, weighing 20 to 25 pounds each, and used by families as an addition to fruit cakes, pies, &c. It is a native of the warm regions of Asia. Heretofore but little attention has been paid to the cultivation of this fruit in Florida, except for variety and ornaments, and it is not usual to observe more than one or two trees in a large garden of several acres in extent, though it is grown here with the greatest ease and perfection, frequently producing fruit weighing 10 pounds, and there is no doubt but that it may be cultivated, preserved, and introduced into our home markets as an article of commerce with great profit to the producer. There is no other variety of this species so easily propagated, and none more hardy, or that yields its fruit so quickly, or produces more abundantly; and the fact that both the fruit and the sugar for preserving it are produced in the same field, with equal facility, gives to the American cultivator a great advantage over the foreign producer in our markets. The citron prepared and preserved by private families in Florida for home use is of much finer quality, lighter colored, and more transparent than the imported. The cost of preparing this fruit for market on a large scale need not be great, and the combination of two articles, green citron and sugar, the cost of producing which does not exceed one-half their actual value, where the two are combined, must leave a large margin of profit to those who engage intelligently and with proper facilities in the business of cultivating and preparing this article for market.

The shaddock (*Citrus decumana*), a native of India or China, is now cultivated in all warm climates, and is called *Arancio massino* by the Italians, *Oranger pampelmouse* by the French, and sometimes in this country mock-orange or forbidden fruit. It was brought from China to the West Indies by Captain Shaddock, from whom it derives its present name. There are at least six varieties, only one of which is useful or desirable as a fruit. Some of these attain a very large size, frequently weighing 10 to 14 pounds. It is chiefly used for ornament or show, and where several sorts of oranges are presented at desert, it forms a striking addition to the varieties in the way of contrast. The most desirable variety of this fruit is sometimes called grape fruit. It possesses a red-

dish pulp, with most agreeable subacid sweetness, and is excellent for
quenching thirst; and from the thickness of its rind will keep longer
than the fruit of any other of the citrus family. This variety is well
worth cultivating for the excellence of its solid vinous pulp, which furnishes a substitute for other acid fruits in pies, tarts, jellies, &c.

Loquat (*Eriobotrya japonica*) is known in the South as the Japan plum.
The tree is an evergreen, and grows 10 to 12 feet high, and is desirable
in every Southern garden on account of its hardiness, withstanding a
greater degree of cold than any of the semi-tropical fruits. It ripens
its fruit in February and March, when most other fruits are gone; is a
profuse bearer, and is readily propagated by seeds and cuttings.

The pine apple (*Ananassa sativa*) is grown in some of the gardens in
the northern portions of the State, but requires protection. South of
parallel 28° it is produced in great excellence and perfection, the pines
frequently weighing 9 and 10 pounds each. This fruit is easily propagated from suckers and crowns, the former preferable, however, the
fruit maturing in three to four months after planting the suckers.

Papaw (*Carica papaya*) is sometimes called the bread-fruit tree. It is
a native of South Ameria. This remarkable tree, though not much
cultivated at the present time in Florida, is worthy of great attention,
not only for the excellence of its fruit, but also for its other extraordinary properties. The tree attains a growth of 20 feet in height, and
yields a large supply of fruit in three years from the seed, and should
be in every garden in Florida south of 30° north latitude. It thrives
well and bears profusely at Saint Augustine. The fruit is pear shaped,
of a light yellow color, varying in size from 3 to 5 inches in length and
from 2 to 4 inches in diameter, and is not unlike a very ripe muskmelon
in taste and flavor, though sweeter. It may be pared and sliced and
eaten raw as a desert fruit, or cut into slices and soaked in water till
the milky juice is out, and then boiled and served as a sauce, or by the
addition of lemon or lime juice, it supplies a most excellent substitute
for apple sauce or tart fruit, to which it is scarcely inferior. The juice
of the pulp also forms an excellent cosmetic for removing freckles from
the skin, and the leaves are frequently used, in the French West India
Islands, instead of soap for cleansing linen. Its remarkable medical
properties, however, are most important, as it is the most powerful ver-
mifuge known, a single dose of the milky juice of the unripe fruit, or of
the powdered seeds of the ripe fruit being sufficient to cure the worst
cases, and extirpate every worm from the system of the patient.

The most extraordinary property of the papaw tree is that related
by Dr. Browne, in his Natural History of Jamaica, in which he says
that the toughest meat or poultry may be made perfectly tender for
cooking, by steeping for eight or ten minutes in the milky juice of this
tree. Dr. Holden, who witnessed its effects in the island of Barbadoes,
says, in the third volume of the Wernerian Society's Memoirs, that the
juice of this tree causes a separation of the muscular fiber in meats that

have been immersed therein and that the vapor of the tree serves the same purpose, it being a common custom with the inhabitants to suspend joints of meat, poultry, &c., in the upper branches of the trees to soften and prepare them for cooking.

Thompson, in his System of Chemistry, makes an extract from a French work on chemistry, entitled *Annales de Chimie*, which states that—

Fibrine had been previously supposed to belong exclusively to the animal kingdom but this tree had been found to contain this substance.

The papaw tree is a perpetual bearer of fruits and flowers, or blossoms, and yields enormous quantities of fruit, a single tree supplying enough for a large family.

Custard apple (*Anona reticulata*) is sometimes called sugar apple. There are upwards of forty varieties of this fruit, and nearly all the species are edible. Almost every tropical country lays claim to its own favorite variety. In Peru it is greatly esteemed, and considered not inferior to any other fruit in the world. The species derives its English name (custard apple) from the consistence of the pulp of the fruit; and its rich color, fragrant odor, and handsome appearance are well characterized in the expression, "apples of gold in pictures of silver."

The Spanish-American cherimoyer (*Anona cherimolia*), and the West India soursop (*Anona muricata*), sweetsop (*Anona squamosa*), and alligator apple (*Anona palustris*) are of this genus. This delicious fruit is produced in excellent perfection as far north as Saint Augustine, and is easily propagated from seed.

FIGS.

Figs are easily raised from cuttings, and begin to bear in two years, producing one good and one or two additional but inferior crops annually. Two hundred trees may be set at nominal cost on an acre. The remarkable vigor and thrift attending the growth of the fig in this State, and the many facilities afforded for an unlimited business growing out of its cultivation and preparation for market, are so decided, that this fruit is worthy, like the orange and cane, of special attention here. A simple preparation of figs by boiling in sirup will furnish a most palatable and wholesome preserve, that only needs to be known to become a universal favorite; and if figs can be prepared for a lucrative market, by drying, anywhere on earth, it can be done in Florida. The London Encyclopedia mentions fifty-six species, of which the following are the most remarkable:

F. carica, the common fig tree with an upright stem branching fifteen or twenty feet high, and garnished with large palinated or hand-shaped leaves. Of this there are many varieties, as the common fig tree, with large, oblong, dark purplish blue fruit, which ripens in August either on standards or walls, and of which it carries a great quantity; the brown or chestnut fig, a large, globular, chestnut-colored fruit having a purplish delicious pulp; the black Ischia fig, a middle-sized, shortish, flat-crowned, blackish fruit, having a bright pulp; the green Ischia fig, a large, oblong, globular-headed, greenish fruit, slightly stained by the pulp to a reddish-brown color; the brown Ischia fig, a small, pyramidal, brownish-yellow fruit, having a purplish rich pulp; the Malta fig, a small, flat-topped, brown fruit; the round brown Naples fig, a globular, middle-sized, light-brown fruit, and brownish pulp; the long, brown Naples fig, a long, dark-brown fruit, having a reddish pulp; the great blue fig, a large blue fruit, having a fine red pulp; the black Genoa fig, a large pear-shaped, black-colored fruit, with a bright-red pulp. It may be propagated either by suckers arising from roots, by layers, or by cuttings. The suckers are to be taken off as low down as possible; trim off any ragged part at bottom, leaving the top entire, especially if for standards, and plant them in nursery lines at two or three feet distance, or they may at once be planted where they are to remain.

The best season for propagating these trees by layers is in autumn; but it may be also done any time from October to March or April. Choose the young pliable lower shoots from the fruitful branches; lay them in the usual way, covering the body of the layers 3 or 4 inches deep in the ground, keeping the top entire, and as upright as possible; and they will be rooted and fit to separate from the parent in autumn, when they may be planted either in the nursery or where they are to remain. The time for propagating by cuttings is either at the fall of the leaf or in February. Choose well-ripened shoots of the preceding summer, short and of robust growth, from about twelve to fifteen inches long, having an inch or two of the two years' wood at their base, the tips left entire; plant them six or eight inches deep, in a bed or border of good earth, in rows two feet asunder.

GUAIAVA.

The name guaiava (*Psidium guaiava*) is a corruption of the Spanish word *guayaba*. Of this fruit there are seventeen different species. It is an evergreen tree or shrub, and indigenous to Brazil, Spanish America, and the West Indies. It is propagated by cuttings and seed, and is sometimes liable to injury from severe frosts north of 28° north latitude, but south of that line it is ever-bearing, yielding its delicious, aromatic, and wholesome fruit all the year round. Only three or four varieties are known and cultivated in Florida.

In the island of Cuba and in Brazil the varieties produced are more numerous, and large quantities of the fruit are made into jellies for exportation to all parts of the world. The fruit of the common guaiava is pear-shaped, of the size of a large hen's egg, and sometimes larger, and has a smooth, pale-yellow skin, inclosing a many-seeded pulp of delicious acidity. In some varieties the pulp is of a light-cream, and others a pale-reddish color. This fruit is greatly esteemed wherever known, and, being slightly astringent, as well as mucilaginous, is very beneficial in bowel complaints. The roots and leaves are also astringent, and are regarded as excellent for strengthening the stomach and bowels. The plant is propagated by seeds, cuttings, and suckers.

POMEGRANATE.

The pomegranate (*Punica granatum*), a shrub or bush-like tree, is a native of Persia and Syria, and grows wild in those countries. It is perfectly hardy in all parts of Florida, and as far north as Hilton Head, S. C., and is widely cultivated and much esteemed in this State for the excellence of its fruit, as well as for the medicinal proper-ties of the rind and the flowers, which are not only an excellent febri-fuge, but powerful astringents, and often used with great benefit in cases of diarrhea. The pulp of the fruit is a delicious sub-acid sub-stance, similar in taste and flavor to the red currant, and is excellent for allaying heat and quenching thirst, and is gently laxative. The fruit of the pomegranate is spherical, the size of an orange, with a gourd-like shell or rind, which is filled with seeds inclosed in membranous cells, and surrounded with a juicy, reddish pulp. There are several varieties of this fruit, comprising early, medium, and late. The early and the medium varieties have a pale yellow skin or rind, with a beau-tiful tinge of red upon the side or cheek, and are sparsely dotted with fine pippin-like spots. The latter sorts have a dark russet-colored rind, and the seeds are of a pale pink color. This tree bears a beautiful urn-shaped scarlet flower; and there is no tree more showy than the pome-granate when in flower. The fruit begins to ripen at Saint Augustine, Fla., about the middle of July, and continues until the middle of Decem-ber. It bears transportation well on account of its hard rind, keeps for several weeks after it has been taken from the tree, and no doubt may be made a profitable market fruit. It is increased by cuttings, layers, and suckers, and thrifty wood two years old strikes fibers as readily as younger wood.

BANANA.

Banana (*Musa paradisiaca*).—Of the banana and plantain (*Musa sapien-tum*) there are several species. They are increased by suckers, and require a rich, moist soil, with warm exposure. Some varieties of these plants are successfully cultivated as far north as 30° north latitude. The best variety for cultivation north of 28° north latitude is the one known as *Musa paradisiaca cavendishii*. This is the most hardy, and seldom attains a height above 8 feet, while the more tender kinds often grow 20 feet high. When the plant is fruiting, and all the flowers are set, it is advisable to cut off the spadix an inch or two above the last tier of per-fectly-formed fruit, in order to hasten and perfect the remaining fruit.

There are few more excellent or delicious dessert fruits than the banana, and as a food plant its importance and value, as compared with other food plants, can hardly be overestimated. In an economical point of view it has never been appreciated in Florida, where but little attention has been given to its cultivation. When it is realized that a plantation of bananas once established has never to be renewed, and that one acre of this fruit will produce as much food as 130 acres of wheat, or 45 acres of potatoes, its value and importance will be readily acknowledged. As this plant is a great feeder, and when once planted lasts for a lifetime, it is of the utmost importance that plantings should be made upon strong rich soil, or that the plants be kept highly manured, to secure permanent supplies of the best fruit. In Brazil and other tropical countries plantations are formed by setting the plants 20 feet apart; but as the kinds usually planted in those countries are of a larger species than those recommended for Northern Florida, plants of the *Musa paradisiaca cavendishii* variety should be set 10 feet apart each way, and in a good soil they will soon cover the ground, as they increase rapidly under favorable circumstances. Each plant produces one, and only one, bunch of bananas, when it is cut down with a sharp spade or ax to give place to succeeding plants.

When the enormous yield of this fruit is considered, and the fact taken into account that when once properly planted it needs no other attention than simply gathering the fruit, the demand for which is almost unlimited, it is evident that its cultivation could be made very profitable.

DATE PALM.

The date palm (*Phœnix dactylifera*) is an excellent and valuable fruit, and is cultivated with entire success south of 28° north latitude, and the tree often perfects its fruit as far north as 30° north latitude. Numerous large and beautiful specimens of this tree may be seen in the gardens at Saint Augustine. It is one of the most beautiful trees of the vegetable kingdom. Its long, graceful, ever-verdant, ever-waving, ever-changing branches make it the most picturesque of all others for landscape gardening, and should adorn the grounds of every homestead in Florida.

The fruit is greatly and justly esteemed by the inhabitants of Egypt, Arabia, and Persia on account of its concentrated and nutritious properties; large numbers subsist almost entirely upon it. It is generally the sole food of the Arabs and their camels in their long and tedious journeys over the desert, the voyagers feeding upon the fruit and the animals upon the stones. The inhabitants of these countries also boast of the medicinal qualities of the date fruit, and of the numerous uses to which the different productions of this tree may be applied. From the leaves they make couches, baskets, bags, mats, and brushes; from the branches or stalks, cages for their poultry and fences for their gardens; from the fibers of the trunk, thread, ropes, and rigging; from the sap, a spirituous liquor, and the body of the tree furnishes fuel. The date palm is propagated from the seeds and suckers, but more successfully from the former. The cultivation of this fruit should be greatly extended, as it may become an important and profitable resource of the inhabitants of Southern Florida. The bunches or clusters of this fruit often attain a weight of 15 pounds.

PERSIMMON.

The persimmon is found wild in every section of the State. The fruit, at least to the natives, is agreeable to the taste, and, ripe or dry, is used largely for the table and for home-made beer. Some Japan varieties are now being introduced, which are said to be of very large size, and seedless. The Japanese esteem the persimmon as their most valuable fruit.

GRAPES.

The grape is found wild in the forests of Florida, and grows luxuriantly. The foreign varieties grow and fruit well, but are afflicted with the *Phylloxera*, and their successful cultivation is not permanently reliable. The Saint Augustine grape, so called, is a choice grape for the table, or for wines. It grows luxuriantly; but is subject to rot during the rainy season. It resembles the grape from which the Madeira wines are made. The Scuppernong, in all its varieties, is the most reliable, as it is a rank grower, prolific in fruit, and free from disease. It makes a wine equal to the best California, and can be grown to great profit. Forty-five feet square are allowed to each vine, and it soon occupies this space, yielding many bushels to each plant. It is a late grape, and a good table fruit. An acre yields upwards of 2,000 gallons of wine.

PEACHES.

The peach is a sure tree here, bearing in two years from the seed, and early varieties of good size and flavor ripening in May, June, and July. The apricot and nectarine are also safe to cultivate. As yet, no disease has affected the trees, and they retain their vigor and prolific bearing for many years.

The northern grown trees do not do well, as they do not seem able to adapt themselves to the climate, but seedlings succeed, make a rapid growth, and are true to the variety planted. Some very fine varieties have been produced, and when this fruit shall receive the attention due to its value and importance, it will be found a valuable product for export to northern markets, where it can be placed in advance of the products there, and command good prices.

PEARS.

The northern varieties of the pear do well here, though they grow and bear fruit at uncertain periods. But the Le Conte pear, as it is called, is a southern variety, equal to the Bartlett, free from disease, and prolific to a high degree. It is rapidly gaining favor, and its cultivation is being extended. It brings from $6 to $10 per bushel in the northern market, where it is placed in advance of the more northern varieties. It yields an enormous profit, greater than the orange in many localities.

APPLES.

The cultivation of the apple here is of doubtful utility, though it is believed that some of the earlier varieties may be advantageously introduced.

NECTARINE, APRICOT, AND PLUM.

The nectarine (*Amygdalus persica*), the apricot (*Prunus armeniaca*), and the almond (*Amygdalus communis*) are all at home in Florida, and not less vigorous and healthy, but not reliable for fruiting in all portions of the State, whether from defective culture or adaptability to the variety of soils is not yet determined, as very little attention has been given to them.

The plum and prune (*Prunus domestica*) are also healthy and productive, but not exempt from the ravages of the curculio, so prevalent at the north. All the varieties of the wild plum are indigenous and abundant in every part of the State. Many of the varieties are of excellent quality, and, when cooked, form a delicious preserve for family use, or for canning.

OLIVE.

With the exception of a few trees grown for ornament, this most valuable tree, the olive, has not been cultivated in this State. That it will succeed well here is evident from the specimens now growing. Recently, attention has been directed to its cultivation, and it will become widely planted. It commences to bear at about ten years from the seed, increasing yearly to the age of thirty years, bearing annually. They are very long-lived; some trees in Europe are known to be eight hundred years old, and show no signs of decay. The fruit and oil are valuable as food and of commercial importance.

NUTS.

The pecan nut and the Madeira nut succeed well, and produce abundantly in the northern portion of the State.

The cocoa-nut and the Brazil nut are produced in the southern portion, and the former is receiving special attention in Monroe County, where thousands are being planted in the vicinity of Key West. Large groves are also being set out on Lake Worth.

TEA.

The tea plant has long been successfully cultivated in Georgia, and through the instrumentality of the Department of Agriculture, at Washington, it has been largely diffused through this State; and while it may not, in the face of competition with foreign labor, immediately become remunerative, yet it will be produced to some extent for home consumption, and in time may become a profitable product.

COFFEE.

The coffee plant has been successfully introduced from South America, and its production in South Florida gives the assurance that it may be grown in that portion of the State at least, if not further north, and that in time it may become an important factor in the varied industries and products of the South. The first pound of coffee ever produced in the United States by open-air culture was grown in Manatee, and received the premium offered by the Department of Agriculture.

STRAWBERRIES.

This queen of small fruits nowhere in the world finds a better location for culture; plants put out in September fruit often in January, frequently in February, and may be counted in full bearing and ripening in March and April. The growers about Jacksonville and up the Saint John's River are many, and shipments have been made largely and profitably. In size, color, bouquet, and taste, they are superior to most, equal to the best, and surpassed by none; the best varieties only are grown. The cultivators pick carefully, select and pack honestly; and Florida strawberries, like Florida oranges, have earned a name. By using refrigerators the fruit reaches New York and the Northern cities, fresh and cool, only about four days from picking. Being always in advance of any other locality by some weeks, the first shipments bring large prices, and the demand keeps pace with the supply.

BLACKBERRIES, HUCKLEBERRIES.

The low-creeping blackberry, or dewberry, abounds in old fields and road-sides, and ripens in April. The high bush, also found in same localities, ripens in June and July; the huckleberry about the same time. All bear well, and can be had for picking. The improved kinds do well where tried.

MELONS.

Watermelons, muskmelons, and cantaleups are readily produced everywhere in the State, and the first are grown extensively for export to the Northern cities. One thousand melons to the acre is considered a fair crop, and the standard size for export is 20 pounds and upwards. They are grown to the size of 60 pounds or more, and no better fruit is produced anywhere.

VEGETABLES.

Along the navigable streams and the lines of railway, the raising of vegetables for export has become an important element in the prosperity of Florida, and is rapidly assuming proportions which claim attention from the transportation lines and the cities of the North. Tomatoes, cucumbers, cabbages, celery, lettuce, beets, turnips, onions, Irish potatoes, snap-beans, Lima beans, pease, egg-plants, okra, and all the variety of vegetables are produced most successfully during the winter and spring months, and as improved methods and better culture are introduced, the results prove that every month in the year may yield handsome returns to the cultivator of the soil, and demonstrate that the cheap lands of Florida may afford greater returns in value for the same labor than the best lands of the North and West.

WORK OF THE SEASONS.

The farm and garden work for the year is briefly indicated, as follows:

In *January* plant Irish potatoes, pease, beets, turnips, cabbage, and all hardy or semi-hardy vegetables; make hot-beds for pushing the more tender plants, such as melons, tomatoes, okra, egg-plants, &c.; set out fruit and other trees and shrubbery.

February.—Keep planting for a succession, same as in January; in addition, plant vines of all kinds, shrubbery, and fruit trees of all kinds, especially of the citrus family, snap-beans, corn; bed sweet potatoes for draws and slips. Oats may also be still sown, as they are in previous months.

March.—Corn, oats, and planting of February may be continued; transplant tomatoes, egg-plants, melons, beans, and vines of all kinds; mulberries and blackberries are now ripening.

April.—Plant as in March, except Irish potatoes, kohl rabi, turnips; continue to transplant tomatoes, okra, egg-plants; sow millet, corn, cow peas, for fodder; plant the butter-bean, lady pease; dig Irish potatoes. Onions, beets, and usual early vegetables should be plenty for table.

May.—Plant sweet potatoes for draws in beds; continue planting corn for table; snap-beans, pease, and cucumbers ought to be well forward for use; continue planting okra, egg-plants, pepper, and butter beans.

June.—The heavy planting of sweet potatoes and cow peas is now in order; Irish potatoes, tomatoes, and a great variety of table vegetables are now ready, as also plums, early peaches, and grapes.

July.—Sweet potatoes and cow peas are safe to plant, the rainy season being favorable; grapes, peaches, and figs are in full season. Orange trees may be set out if the season is wet.

August.—Finish up planting sweet potatoes and cow peas; sow cabbage, cauliflower, turnips for fall planting; plant kohl rabi and rutabagas; transplant orange trees and bud; last of month plant a few Irish potatoes and beans.

September.—Now is the time to commence for the true winter garden, the garden which is commenced in the North in April and May. Plant the whole range of vegetables except sweet potatoes; set out asparagus, onion sets, and strawberry plants.

October.—Plant same as last month; put in garden peas; set out cabbage plants; dig sweet potatoes; sow oats, rye, &c.

November.—A good month for garden; continue to plant and transplant, same as for October; sow oats, barley, and rye for winter pasturage or crops; dig sweet potatoes; house or bank them; make sugar and sirup.

December.—Clear up generally; fence, ditch, manure, and sow and plant hardy vegetables; plant, set out orange trees, fruit trees, and shrubbery; keep a sharp lookout for an occasional frost; a slight protection will prevent injury.

TIMBER AND LUMBER.

Of the States, Florida has the largest area of original growth of timber. Excluding land in cultivation, the area covered by lakes, rivers, savannas, &c., there are probably nearly, if not quite, 30,000,000 acres of land covered with timber, and of this the yellow pine is fully three-quarters. The level lands, rolling lands, are mostly covered with the yellow and pitch pine, which attains a great size in girth and length. The lower lands, near rivers, lakes, swamps, abound in valuable timber, of which live oak, other species of oak, hickory, ash, beech, cedar, magnolia, sweet bay, gum, cypress, constitute a great proportion. The red cedar is particularly adapted for lead pencils, and largely exported to Europe for the best manufacturers, as also North and East. The magnolia and bay are fine woods for ornamental furniture; the cypress valuable for shingles, sash, doors, blinds and inside finish, railroad ties. The yellow and pitch pine has a world-wide reputation as being the best for any and all uses where strength, elasticity, and durability are desired, and is now being largely used in ornamental and expensive structures. Finished up in its natural grain for inside work, floors, frames, pillars, arches, roofs, it presents that substantial as well as rich finish not attained with other material. An extensive business has long been prosecuted in the western portion of the State in the exportation of pine timber and lumber, and live oak and cedar have been supplied in large quantities for naval architecture and the manufacture of pencils. Large establishments for the manufacture of lumber for exportation to northern and foreign ports are located on the navigable streams and railroads. Naval stores are also largely exported, and afford a reliable source of revenue to the State and the manufacturers.

THE FORESTS OF FLORIDA.

From a gentleman who has been connected with the Tenth Census of the United States the following information has been obtained relative to the forest growth of this State:

The variety of trees in Florida is greater than in any other of the States, there being nearly, or quite, two hundred arborescent species, or half of all in the United States.

One of these, the yellow pine, is a hundredfold more abundant and more valuable than all the others combined. This is the most valuable of American trees for framing-timber, flooring, &c., and the present reserve of it is almost wholly confined to South Georgia and the northern half of Florida. This is the tree from which turpentine and rosin are obtained. There is very little yellow pine suitable for shipping south of latitude 29°, but plenty for home consumption as far south as latitude 27°.

South of latitude 27°, to the southern limit of the State, there is plenty of the Southern pitch pine, which is equally good for turpentine, and furnishes very good building material. There is also a belt of it, perhaps ten miles in width, bordering the entire coast. There are five other kinds of pine in the State, of less value and in small quantity.

Next to the yellow pine in abundance and value rank the bald cypress and live oak. Both are found throughout the State in abundance and of large size. The qualities of live oak are well known, better than those of the cypress, a timber only second in value to yellow pine, and for durability and lightness superior to it. Several mills are now sawing cypress, and the manufacture of it must increase.

The red cedar of Florida is celebrated as being the only wood suitable for pencils. The demand for it is so great that the supply will probably be exhausted before many years.

Throughout the pine woods, north of latitude 27°, are scattered much post and black-jack oak, which furnish superior fuel. The other kinds of trees occur in groves or hammocks and border water-courses.

Along the southern half of the Saint John's and Indian Rivers the palmetto is the most abundant tree, but, as a rule, the hammocks are composed in most part of the live oak and the water or willow oak, with much red bay, magnolia, and hickory. Besides these, there is a great variety of smaller trees, with handsome evergreen foliage and wood suitable for various purposes. In some hammocks the sour and bitter-sweet oranges abound.

The trees most noted for beauty of wood are the "curly" variety of the yellow pine, the red or sweet bay (sometimes called Florida mahogany), the magnolia, black cherry, holly, &c.

In Jackson and Gadsden Counties are found the black walnut, sour wood, bass wood, beech, birch, sugar maple, cottonwood, sycamore, and many other Northern trees. This section is not suited to the growth of the orange.

Trees common in low hammocks and swamps are the white bay, tan bay, elm, ash, hickory, red maple, sweet gum, sour gum, poplar, hackberry, iron-wood, &c.

The number of species of the most important genera are as follows: Oak, 13; pine 7; hickory, 5; elm, ash, maple, magnolia, and gum, 3 each.

South of the Caloosahatchie River, and on the east and west coasts, as far north as Mosquito Inlet and Cedar Keys, the trees are nearly all of subtropical species, among which are some of great beauty and value, such as the mahogany, lignum vitæ, prince-wood, mastic, wild tamarind, calabash, royal palm, Indian fig, crab-wood, &c. But these are found in too limited quantity to be of any practical importance.

In addition to these are the Madeira, frankincense, white-wood, cork-wood, sea grape, green ebony, wild cherry, buttonwood, black and rock mangrove, and numerous others.

A variety of the caoutchouc grows in South Florida, which furnishes a gum possessing the qualities of the India rubber of commerce.

It is believed that the cinchona, which furnishes that valuable article quinine, could be successfully cultivated. Other trees and plants of great merit and utility can doubtless also be introduced.

An experimental garden, under the auspices of the Department of Agriculture, would be attended with signal benefit to the interests of the country.

LAKE OKEECHOBEE AND THE EVERGLADES.

But little is known of the vast region covered by Lake Okeechobee and the everglades. Much of that immense area is unexplored and un-

surveyed, and few, besides the wandering cowboys, ever traverse its
wild solitudes. Its almost absolute inaccessibility has practically shut
it out from settlement; and for years it has been occupied by the small
remnant of the Indian tribes that once owned the country, and by vast
herds of cattle, which thrive and fatten upon the rich pasturage.

Very recently the eyes of capitalists have been directed to this remote
wilderness, and science has demonstrated that this section can be reclaimed and subjected to cultivation. A careful analysis of the soil re-
veals the fact that it contains a remarkable percentage of the elements
necessary to sustain vegetation, and that the most exhausting crops can
be produced for years without any diminution of its natural fertility.
These lands, when drained and properly handled, will be found capable
of supplying a very large portion of the sugar now consumed in the
United States. They are admirably adapted to the production of this
important staple, and are fully equal, if not superior, to the most valuable lands devoted to this crop in Cuba and Louisiana. Here the cane
reaches maturity, and regularly goes to tassel, showing that the stalk
has reached its highest point of development, and that its juices contain the largest amount possible of saccharine matter. One planting of
the seed suffices for several years, the plant *rattooning* for a number of
seasons from the old stubble. Another advantage of the utmost consequence is found in the fact that, situated as this region is, below the
frost line, the cane can be cut and ground to suit the convenience of the
planter.

Mr. H. A. Hough, who resides on Twelve-Mile Branch, an arm of the
Caloosahatchee River, had his sugar-mill destroyed by fire in January,
1881, before his cane had been harvested. The building was re-erected
in the following April, when the cane was cut and ground, making a
full average yield of superior sugar and sirup, for which the highest
prices were obtained. It is customary in that section to replant cane
but once in seven years. The planter is never harassed by fears of
having his crop injured or destroyed by the sudden advent of cold
weather.

Besides cane and rice, tobacco, cotton, jute, indigo, and the whole
series of tropical and semi-tropical fruits can be successfully and profitably cultivated.

The soils and situation of this region are such that the entire range
of vegetables can be put into Northern markets before the truckers
around Boston, New York, and Chicago have commenced to break their
ground for the reception of seed. In this respect this region has no
rivals in this country, and can compete with the Bermudas and Bahamas.
The increase of wealth and the growth of luxury in the United States
have created a steady and growing demand for table dainties, and the
prices paid for fruits and vegetables out of their usual season are largely
remunerative. Tomatoes, English pease, string beans, egg-plants, okra,
and other garden products can be shipped from this section during the

entire winter—from December to April. In the cultivation of such articles there is a mine of wealth opened to the enterprising and industrious settler. But little heavy clothing, and scarcely any fuel, except for culinary purposes, is required in this genial climate, so that there is a great saving in these items, so costly in a more rigorous latitude.

The question of transportation, which has heretofore been a barrier to the occupation of this desirable country, is about to have an early and satisfactory solution. By the time this report will be given to the public a line of steamers will be in operation upon the Kissimmee River, connecting Lake Okeechobee with the new city of Kissimmee, recently laid out on Lake Tahopekaliga, in Orange County. Here a junction is effected with the South Florida Railroad, a line stretching from Sanford, on the Saint John's River, to Tampa and Charlotte Harbor. Communication will also be had with the West as soon as the Caloosahatchee River is open to navigation.

A company was chartered by special act of the legislature of Florida, March 8, 1881, for the purpose of purchasing and improving certain tracts of land in Florida, the building of canals and other lines of transportation, and the carrying on of all other business incidental thereto.

The following information is taken from a pamphlet recently published by the above company:

This company has a concession from the board of internal improvement of the State of Florida for the reclamation of all the lands lying south of townships 24 and east of Peace Creek, this area containing upwards of 8,000,000 acres.

The United States survey, made in 1879 by Col. J. L. Meigs, established the elevation of Lake Hickpochee, adjoining Lake Okeechobee, as being 22 feet above mean low tide, and he recommended the construction of a drainage canal similar to that now proposed to be established. These surveys and observations have recently been verified by a corps of engineers in the employ of this company, who found Lake Okeechobee to be 25 feet above tide water.

Lake Okeechobee, situate about the center of this 8,000,000 acre tract, is upwards of 40 miles in length by 25 miles in width, or covering an area of over 1,000 square miles. It has an outlet, but receives the drainage of a number of lakes intercepted by the Kissimmee River, also the waters of Fish Eating, Taylor's, and Mosquito Creeks, which vary from 20 to 150 feet in width. During very heavy falls of rain, this lake rises to such a point as to not only overflow its banks, but to cause the waters of the rivers to be backed up, so that the country becomes more or less submerged, until the waters find the ocean and gulf through the tortuous and inefficient channels of widely-separated streams.

It is proposed to provide against these periodical overflows by the opening of canals from Okeechobee to the Saint Lucie and Caloosahatchee Rivers that will not only permanently lower the level of the lake, but at all times furnish a safe outlet to the gulf and ocean for the waters of the lake and confluent streams, and which will also afford means of transportation for the products of the Kissimmee Valley and surrounding country.

The reclamation of the land included in that portion of the peninsula of Florida south of latitude 28° 15' N., and generally east of Pease Creek, embraced in the counties of Monroe (5,000 square miles), Dade (5,000 square miles), Brevard (4,000 square miles), and portions of Manatee (5,000 square miles), and Polk (1,900 square miles), is a problem the magnitude of which can be more readily comprehended when we consider that the territory in question covers 1,000 square miles in excess of the combined area of the States of Rhode Island (1,300 square miles), Connecticut (4,700 square miles), New Jersey (8,300 square miles), and Delaware (2,120 square miles). In other words, over 17,000 square miles of the most tropical portion of the most tropical State in the Union are to-day ready to respond to an intelligent, systematic, and properly directed effort towards placing them in the category of tillable and available acres, embracing no barren prairies nor mountain wastes. There are but few acres not susceptible to a high degree of cultivation. Lands which in a more northern climate would be regarded as valueless will here yield luxuriant and remunerative crops. For example, the scrub palmetto or poorest pine barrens of Southern Florida will produce, without fertilizers, large crops of Sisal hemp, yielding a profit to the acre which compares favorably with the returns from the richest land when cultivated in sugar tobacco, or cotton. The same character of land will produce from 50 to 75 bushels of upland rice to the acre—a three months' crop; or at a trifling original outlay, 15,000 pineapple slips, set to the acre, will, from the poorest scrub land, yield a return far in excess of the brightest dream of the Northern farmer. Other valuable tropical products adapted to these lands could be mentioned, which, in a more northern climate, would yield nothing to agriculture. This glance at the possibilities to be realized from the cultivation of third-rate pine and stunted "Black Jack" lands prepares us somewhat for a better appreciation of the capabilities of the soil designated as "rich lands," and named in the following order : First, "swamp lands"; second, "low hammock"; third, "high hammock"; and fourth, "first rate pine, oak, and hickory lands." It will only be necessary to call attention to the fact that the "swamp"— or lands subject to overflow—are intrinsically the most valuable lands in Florida. To adapt them for successful cultivation a systematic plan for their drainage will be indispensable; when thus prepared their inexhaustible fertility sustains a succession of the most exhaustive crops with astonishing vigor. The greatest yield of sugar ever realized in Florida (4 hogsheads per acre) was produced on this description of land.

It will be impossible to form or convey an adequate idea of the importance and extent of this enterprise, developing, as a consequence, a new and vast territory unlimited in resources, and of such material and varied agricultural wealth as can be furnished by no other State in the Union; opening to cultivation a tract of sugar lands the soil of which is identical to that of Cuba and Louisiana of a productive power apparently inexhaustible and unequaled in area by any country on the globe The prominent natural requisites to the growth and maturity of the sugar cane under the most favorable conditions, obtain here in a marked degree.

A moderate proximity of these lands to the sea and gulf, a dry, warm spring, showers during the afternoons in June, July, and August, followed by a comparatively dry autumn, a condition necessary for converting the starch into saccharine matter, are characteristics of the peninsula of Florida south of the 28th parallel of latitude. The importance of this one crop as affecting the material wealth of our country can be more readily comprehended by a bare comparison with the enormous output in precious metals from our western mines, those great store-houses of national wealth. The import duties on sugar for manufacturing purposes from the year 1847 to 1879 varied from 2¼ to 4 cents per pound. We paid out for sugar and allied products during this period $1,800,000,000. Our western mines produced $1,700,000,000, or, in other words, during a period of thirty-two years as a nation we paid out in round numbers $100,000,000 in excess of the total output in bullion of our famed bonanzas of the west for an article of consumption every pound of which could have been produced from the soil of Southern Florida.

The choice sugar lands of Louisiana are rated at from $100 to $150 per acre, similar in character to those just described, which mature the cane to perfection, and are located below the frost line.

The terms of the contract with the board of internal improvement of the State of Florida give to this company one-half of all the land reclaimed by the lowering of the waters of Lake Okeechobee.

The peculiar characteristics of the coraline foundation upon which the peninsula of Florida has, by a gradual and cumulative process, been raised to its present level above the waters of the ocean; the configuration of its surface and that other marked geographical feature as indicated by the enormous extent of her coast line, exceeding 1,100 miles on the Gulf and Atlantic, indented by numerous large bays and estuaries; the uniform width of the lower portion of the peninsula and comparative short distance separating the waters of the Gulf and Atlantic; taken in connection with the successive slight ridges or table lands, generally parallel with the coast line, comprehending within their borders long reaches of savanna, prairie, and marsh, and increasing in altitude as we proceed towards the interior or water-shed of the Kissimmee River, whose flows empties into that grand island reservoir, Lake Okeechobee—we have before us the necessary data upon which to develop the plan for the solution of the problem of successfully draining and reclaiming this vast territory of notably rich lands.

An analysis of the soil taken from the saw-grass marsh on the border of Lake Okeechobee, by the distinguished Dr. Rogers, professor of chemistry in University of Pennsylvania, gives the following results:

Organic matter (vegetable mold)	55.00
Silica	21.80
Carbonate of lime	21.50
Iron, a trace	0.00
Water, not volatilized at 212°	1.70
	100.00

As the operations now in progress promise to add an immense area, comprehending no less than 18,000 square miles of the very richest and most productive lands to the agricultural resources of the country, all matters connected with this vast enterprise contain information of the greatest importance.

Col. Ingham Coryell, general superintendent of the Atlantic and Gulf Coast Canal and Okeechobee Land Company, furnishes the following information relative to the plans and points of operation for the proposed drainage of Okeechobee:

We have been at work for the past two months, under the supervision and direction of Capt. F. A. Hendry, in cutting a channel from the headwaters of the Caloosahatchee to Lake Hickpochee, and have succeeded in opening one 20 feet wide, and now having a depth of 3 feet. His report is that it is daily deepening, and so soon as the waters lower in Okeechobee, what is now a marsh overflow will be concentrated from Hickpochee in the channel as now made. It will cut very fast, and thereby enable the dredge-boat, nearly finished, at Cedar Key, on its arrival over the Caloosahatchee, to be forced up and over the 16-foot elevation into Hickpochee. When there we are 3 miles from Okeechobee, with only 2½ feet elevation to overcome to get into that lake, which is entirely practicable. By taking the dredge by a circuitous route, throwing our excavation in our rear as we proceed, we dam up our passage *en route*, as far as soft mud and vegetable growth will obstruct the passage of water.

It is not our intention to open a flood on the valley of the Caloosahatchee, and we will use every means to prevent it. Our objective point of operations will be on the southeast end of Lake Okeechobee, where our first cuttings will commence. We are now building a second dredge on Lake Tohopokaliga, which is the head of the Kissimmee River, in Orange County, at the newly laid-out city of Kissimmee, and a short cut of 2 miles at the southwest corner of that lake into Reedy Creek will afford us plenty of water; thence into the Kissimmee River down to Okeechobee, where we hope to join the Cedar Key dredge, and carry both to the southeast side of Okeecho-bee. The Cedar Key dredge has a cutting capacity of 22 feet wide and 10 feet depth.

The Tohopokaliga dredge will have a cutting capacity of 25 feet and over, and 10 feet depth. Both dredges have stern wheels and are self-propelling. At the south-east side of Okeechobee there is an open slough from 4 to 6 miles in length, and evi-dently at one time an open river, heading in the lake and emptying into the Atlantic at Miami, New Hillsborough, or some other river now carrying the surplus waters of the Everglades, reaching their head, into the ocean.

Further investigation may cause us to utilize this open slough as one of the points to commence our cutting, which will first be by a cut 4 feet deep and the width of the full capacity of our dredge. The other dredge will commence near by, and will con-verge to a point, say 4 miles from the lake, where they will meet, and to which point other cuts will be made if found necessary, concentrating the water and flows, and the force thus concentrated will possibly cut its own channel from that point southeast to the ocean. It will at least go somewhere on the down grade, and be gotten rid of in a direction where no damage will be done to any one. As to the extent of the open-ings to be made, I don't think it is within the power of engineering science to deter-mine by a survey and calculation. We can calculate the surface level of Okeechobee to the depth of one foot to contain 31,363,200,000 cubic feet of water, and the lowering of the lake one foot reduces its contents fully one-fifth of its entire capacity and from its shallowness will expose or reclaim possibly over one mile of its entire border; hence the cubic feet of water by the next reduction of one foot will be very materially less-ened, and so on by every foot of reduction. A reduction of one foot solves the prob-lem. We have measured the inflow at the mouth of the Kissimmee, when the waters were within its natural banks, but during a flood, its spread is for miles east and west over a saw-grass marsh, and impracticable to measure with any certainty for several reasons, the principal of which is the irregularity of the flow at points through the saw grass, where greater or less obstructions exist. For this reason, no actual calcula-tion can be made of what cuts in area are required, but we are reduced to and forced to accept the general positive requirement of making "the outflow greater than the inflow to prevent overflow." The reduction of the lake one, two, or three feet will form a reservoir that will prevent an overflow during an unusual rainfall, even if the outflow should not be fully equal to an unusual rainfall and inflow, and will prevent an overflow.

One important and encouraging point in success of the enterprise is the fact that we meet with no rock, hard pan, or clay obstruction, and the cutting through the mud formation by the force of water concentrated must be very great, and it is reasonable to suppose will cut and deepen its own channel. Tests by the sounding rod have been made from the Caloosahatchee up to and through Hickpochee, and thence to Okee-chobee, and on the Atlantic side from Lake Worth, for a distance sufficient to know that no rock or hard pan exist to a depth that will reach the bottom level of Okee-chobee. Of this there can be no question. You ask will the lowering of Okeechobee effect the whole of the drainage desired? By no means; the work is then only begun. The straightening of the channel of the Kissimmee must follow, and a series of drain-age operations at other points, which space will not allow me to illustrate.

THE INDIAN RIVER AND LAKE WORTH REGION.

This interesting and attractive portion of the State is now engaging a large share of public attention, in consequence of the certainty that it will shortly be blessed with the advantages of cheap and rapid facilities for transportation. The eastern shore of this river is washed by that remarkable ocean current, the Gulf Stream, giving it at all seasons a uniform temperature, and very effectually protecting it against sudden falls of the thermometer. The orange grown in this region attains a perfection of richness and delicacy of flavor not found elsewhere, and brings the highest prices. Here seems to be the natural home of the pine-apple and cocoanut, and these fruits thrive and flourish in tropical luxuriance. Its waters abound in all the varieties of fish known in these latitudes, and also in the finest oysters, and that prized product of the sea, the green turtle. Here will be established almost certainly extensive fisheries and canneries for the preparation of these delicacies for distant markets, and such industries promise, in the near future, to attain large proportions.

To the devotee of the gun and fishing-rod this country opens a new and fertile field of operations. Game is found in the utmost profusion, and in the winter especially, when the migratory birds seek a milder climate, the abundance of fowls along the course of this river is wonderful. Bears, deer, wild turkeys, and ducks of all varieties abound, and birds of rare and beautiful plumage invite the attention of the taxidermist. When this section is opened, and suitable accommodations are provided, it is destined to become a favorite resort for sportsmen.

The same company that proposes to drain Lake Okeechobee and the everglades has also contracted with the State board of internal improvements to connect the waters of the Saint John's and Indian Rivers by a system of canals. The proper surveys have demonstrated the feasibility of the proposed plan, and it is expected that the work will be commenced at no very distant period. When completed, this enterprise will open 330 miles of safe inland navigation, extending from Lake Worth on the south to the mouth of the Saint John's River on the north, at a point somewhere in the vicinity of Jacksonville.

From the pamphlet already quoted I take the following paragraphs :

This company has acquired by purchase the franchise of the Atlantic Coast Steamboat, Canal, and Improvement Company for the construction of a canal, suitable for commodious light-draft steamboats, commencing at the Saint John's River, extending thence in a southerly direction, connecting Pablo Creek, San Diego, Mantanzas, Halifax, and Hillsborough Rivers, Mosquito Lagoon, Indian River, Saint Lucie Sound, Jupiter River, and Lake Worth, thus affording nearly 330 miles of continuous navigable inland waters, lying adjacent to and generally parallel with the east coast of Florida, being separated from the ocean by peninsulas and extended narrow islands ; these natural barriers guaranteeing a safe and continuous navigation throughout the entire year. The water is salt, being constantly renewed from the inlets at Saint Augustine, Mantanzas, Mosquito Lagoon, Indian River, and Lake Worth. These inland waters affording at present an almost unbroken line of communication, may, at a reasonably moderate expenditure in systematic construction, presenting no embarrassing engineering problems, be developed into a grand canal possessing features peculiarly its own. Merely where the artificial work of joining river to river is performed can it be regarded as a canal proper, as from these points it develops into those majestic arms of the sea, from 30 to 120 miles in length, varying from 1 to 6 miles in width, bordered on either side by a country generally well elevated, enjoying unbounded natural agricultural resources, a semi-tropical luxuriance in beauty of foliage, scenery of an exceedingly variegated and picturesque character, and blessed with a climate throughout the entire year the most equable and salubrious enjoyed by any State in the Union. The soil is generally sandy, with an admixture of disintegrated coral and shell with alluvial and organic matter, capable of supporting successive crops without the aid of manure.

Large bodies of high and low hammock lands of palmetto, oak, bay, hickory, &c., extend along the shores, adjacent to which, on the west shore, are tracts of high open pine lands, bordered by prairie, savanna, and marsh. Experience demonstrates that the soil is not affected by drought in the same degree as other lands, nor does it suffer from too much rain, and, being exceedingly friable, requires no other preparation than grubbing and plowing to adapt it at once for the production of crops covering the widest scope, embracing tropical and semi-tropical fruits and fibrous plants in great variety, and maturing to that degree of perfection developed at no other point within the bounds of the United States.

The topographical features of the country will permit of a general alignment for the canal on the most direct route connecting the several rivers and navigable streams above alluded to. The method of performing the major part of the excavation will be by labor-saving appliances especially designed for this work, combining great efficiency with ease of manipulation and economy in power.

It will be impossible to form or convey an adequate idea of the importance and extent of this enterprise, developing, as a consequence, a new and vast territory, unlimited in resources, and of such material and varied agricultural wealth as can be furnished by no other State in the Union; opening to cultivation a tract of sugar lands, the soil of which is identical to that of Cuba and Louisiana, of a productive power apparently inexhaustible and unequaled in area by any country on the globe. The prominent natural requisites to the growth and maturity of the sugar cane under the most favorable conditions obtain here in a marked degree.

In view of the natural advantages generally cited, experience and statistics guarantee that continued health may be anticipated with as much confidence as in any other section of the country; lands, cheap and readily accessible, easy of tillage, from the fact that, owing to the friable character of the soil, cultivation is neither laborious nor expensive; the harvesting of crops covering the widest scope, embracing nearly all of the grains, fruits, and vegetables of the Northern, Middle, and Southern States; besides tropical and semi-tropical fruits and fibrous plants in great variety, and maturing to that degree of perfection developed at no other point within the bounds of the United States, or indeed anywhere north of Central America—it is fair to assume that immediately subsequent to its completion the unoccupied lands bordering the canal will be entered upon by those experienced in agriculture; capitalists, merchants, speculators, and all of the elements that enter into the thrift and prosperity of a new country will settle among its borders, forming the nucleus of future thriving villages, communities, and cities, constituting the pioneers of that great and inevitable people destined to populate and harvest from the rich prairie, savanna, and upland of the interior bountiful and staple crops, for the production of which nature for past cycles has been preparing the soil by the enriching process of growth and decay of a luxuriant vegetation.

As the cultivation of sugar will probably be the largest and most important industry of this region, it may not be amiss to insert a valuable article upon that subject from the pen of the well-known Dr. C. J. Kenworthy, of Jacksonville, from whose writings copious extracts have been previously taken:

The list of Florida productions is a long one, embracing nearly all the cereals, fruits, and vegetables of the Middle, Northern, and Southern States, as well as the fruits, vegetables, and medicinal and fibrous plants of semi-tropical and tropical countries. Since the settlement of the State the inhabitants have confined themselves to the culture of two or three crops, and old customs cling to the majority as do barnacles to the hull of a stranded vessel in a tropical sea.

Under English rule sugar cane proved a profitable crop, and for many years anterior to 1861 it was extensively and profitably grown by Hon. D. L. Yulee and others. For the successful culture of sugar cane a comparatively dry and warm spring, a high thermal range, coupled with frequent torrential showers, preceded and followed by sunshine during the summer, and a dry and warm fall and winter, are essential. These climatic conditions exist to an eminent degree in Southern Florida, as established by observations taken at Fort Myers, on the Caloosahatchee River, and at Fort Dallas, Cape Florida.

MEAN TEMPERATURE.

Stations.	Years and months.	Spring.	Summer.	Autumn.	Winter.	Year.
Fort Myers	4 0	75°.4	82°.4	77°.0	65°.3	75°.0
Fort Dallas	4 6	74°.7	81°.5	76°.3	68°.6	74°.7

Rainfall is an important factor in the growth and maturation of the cane; and the necessary hyetal conditions exist in the southern portion of this State.

RAINFALL IN INCHES AND HUNDREDTHS.

Stations.	Years.	Spring.	Summer.	Autumn.	Winter.	Year.
Fort Myers	4	11.07	31.61	11.09	8.33	62.91
Fort Peace, Ind. T.	3	1L 13	26.25	16.84	8.76	62.98

With a high thermal range and ample rainfall during the summer months, the cane attains a development rarely excelled in the West Indies. In the southern portion of the State it ratoons and tassels, and attains saccharine maturity. Even in the northern portion of the State it reaches a more perfect growth and development than in a large portion of Louisiana. We have seen it stated that from ten to fifteen ripened joints to a cane is deemed a good yield, and this can be excelled on the high sandy lands of the northern portion of the State. It is admitted that Cuba is the home of the cane, and that climatic conditions are the elements of its success. For the purpose of comparison we will give the mean temperature and rainfall at Havana:

	Years.	Spring.	Summer.	Autumn.	Winter.	Year.
Mean temperature	8	75°.7	84°.2	75°.5	68°.4	75°.9
Rainfall	5	7.5	14.23	11.48	6.54	39.76

From the data quoted, it will be seen that the rainfall at Havana is less than one-half as great as that at Fort Myers, and as a result the cane attains a greater development of stalk and amount of saccharine matter in Southern Florida than in Cuba. Even the mean annual temperature of Havana is but nine-tenths of a degree above that of Fort Myers. But the great advantage Southern Florida possesses over the West Indies is the excessive rainfall during the summer—frequent orrential showers followed by bright sunshine, with a high mean temperature, causing a luxuriant growth of cane that surprises the resident of the tropics.

With a duty of upwards of three cents per pound on imported sugar it is surprising that the culture of cane in Florida has been so long neglected, more especially when we take into consideration the fact that the people of the United States have paid more for sugar and its allied products since 1849 than the value of the precious metals produced by the mines of the Western States and Territories during the same period.

Sugar-cane is extensively cultivated in Louisiana, but the climatic conditions are not as favorable as those of Florida. For the perfect maturation of the cane it is admitted that an annual mean temperature of 75° F. is necessary. As a rule, in the former State the summer rains are insufficient for vigorous growth except in the lowlands; the occurrence of cold rains during the autumn, early frosts, and a low thermal range in the autumn seriously interfere with the vito-chemical action necessary to change the starch into sugar. Taking New Orleans as a point of comparison, we find the temperature and rainfall to be:

	Years.	Spring.	Summer.	Autumn.	Winter.	Year.
Temperature	39	68°. 9	80°. 9	69°. 3	55°. 7	68°. 7
Rainfall	39	11. 8	15. 9	11. 2	12. 3	50. 7

In Southern Florida the cane need not be ground until after Christmas, but to escape early frosts in Louisiana they are compelled to crush it before it is matured. In a New Orleans paper of recent date we find the following: "Many plantations are grinding, but the cane is somewhat too green yet. Estimates based on reports from a large number of plantations promise a yield of about 136,000 hogsheads, a falling off of four-tenths as compared with last year." Even with this diminished yield the State will receive over $15,000,000 for its sugar crop, and these and many more millions should be made to enrich Florida.

In Louisiana it is necessary to cultivate the cane on the low alluvial soils, but owing to the rainfall in this State during the summer the cane will yield large crops on high and even sandy lands. In his work on Florida, published in 1823, Vignoles remarked:

"Respecting sugar, the recent successful trials that have been made upon it have determined the curious fact that it will grow in almost any of the soils of Florida south of the mouth of the Saint John's River. The great length of summer, or period of absolute elevation of the thermometer above the freezing point, allows the cane to ripen much higher than in Louisiana."

From the best information we have been able to obtain, the cane produced in Duval County on elevated lands is larger, longer, and more perfectly ripened than the product of Louisiana. We have examined cane grown on the Indian River which had from forty-six to fifty-four ripened joints. In the beginning of this year Professor MacCauly, of the Smithsonian Institution, visited the Indian camp in the Big Cypress, 60 miles east of Fort Myers. On his return he informed me that the Indians were engaged making sugar, and that the ripened cane-stalks would measure from 18 to 22 feet in length.

Vignoles says: "Perhaps it may be thought that Florida presents but little to tempt the large sugar planter; granted, but it is undoubted, if the culture of the cane should be adopted on a small scale, by the same proportionate number of cultivators that are in the habit of raising cotton in Carolina, Georgia, and Alabama, their labor would be amply repaid and a source of wealth opened; particularly should some public-spirited and enterprising individual establish, at central and eligible points, sugar-mills to receive the small crops, precisely on the same principle that cotton-gins and rice-mills exist in Southern States. This would augment the population and increase the resources of the country sooner and better perhaps than any other mode."

If the establishment of sugar-mills at central and accessible points was desirable in 1823, it is more so in 1881. For years I have advocated the erection of a large sugar mill and refinery in Jacksonville, where cane can be supplied by small growers and converted into sugar and molasses. Land and labor are cheap, taxes are low in the country, firewood at the saw-mills is costless, and instead of consuming sugar grown in Louisiana and Cuba we should produce enough for home consumption and a half million hogsheads for exportation. With a suitable climate and soil, coupled with cheap labor and an opportunity to obtain a supply of cane on reasonable terms, it seems farcical that Floridians should consume sugar produced within one hundred miles of the State; pay the freight from Cuba to New York, an import duty at the rate of on No. 13 to 16 sugar, 3 $\frac{7}{6}$ cents, and on No. 16 to 20, 4 $\frac{1}{6}$ cents per pound; the cost of refining in a Northern city, where labor is high, and where firewood and coal are expensive, coupled with freights, commissions, profits, &c. There may not be "millions in it," but there is a profitable investment for some person who will erect a sugar mill and refinery in Jacksonville. If a market could be found for cane, its production would be insured, and thousands of persons would engage in its cultivation; the young orange grower, small farmer, and the owners of small patches and town lots would plant it, and an ample supply for a mill and refinery could be secured. This city is connected by river or railroad communication with every portion of the State, and with low freights the cane could be shipped from distant points. Or, to avoid the cost of transportation of cane, producers at a distance could press the cane, barrel the juice, add the milk of lime, and ship it to its destination. The superannuated third-class machinery at present used in the State consigns nearly one-half of the juice of the cane to the bagasse heap; hence an annual monetary loss. The central factory system, with perfect machinery, would materially increase the yield of sugar from the cane produced. From the best information before me I have reason to believe that a central mill could afford to pay the producer $4 per ton for the cane, and he would find it a more profitable crop than cotton. A central factory means a divorce of the agricultural part of the sugar production from the manufacturing, mercantile, and financial part, and that it would prove a profitable investment we are assured. In 1858 the growers in Louisiana produced 1,124,592 hogsheads of sugar, worth $120,000,000; and if this can be done in the unsuitable climate of Louisiana, the question arises, what can be done in Florida with her soil and climatic advantages? We might furnish many facts and figures relating to this industry, but space says, hold, enough.

ORANGE CULTURE.

This is one of the industries of Florida that has suddenly attained very considerable proportions. From barely nothing, in a commercial sense, at the close of the war, the business has grown to be worth $1,000,000 in 1880. Measured by the progress of the past, it is destined to become, in a very short time, one of the leading industries of the State. Last year there were exported at least 45,000,000 of oranges. The business so far has been very successful, and is daily inviting more capital and enterprise. There is already $10,000,000 invested in orange groves in the State, with a field open for the profitable employment of $50,000,000 more. Lands suitable for growing oranges are in abundance and at low prices. Orange groves can be found in almost every part of the State, and on all varieties of soil well drained, the groves numbering each from 10 to 10,000 trees. Hardly a family outside of the cities but cultivates a greater or less number of orange trees, and many residing in the cities do the same. Some of the largest groves in the State are owned by persons living in the towns, or by non-residents. In some of the counties there were raised as high as from 4,000,000 to 6,000,000 of oranges last year; and narrow-gauge railroads are rapidly being built to afford the middle counties facilities for shipping their enormous crops to market. Three such roads have been completed within the past few months, and others are projected, while more are under contemplation. Oranges are shipped from off these roads to New York in eighty and ninety hours' time.

Within the past few years orange culture in Florida has also attained great perfection. It has reached that position where it is possible to analyze the cost of production. Abundant evidence exists that can be brought forward to show the value and profit in it for the investment of capital. Results have shown that there is not at present any pursuit, where the tilling of the soil is involved, that will yield larger returns with less fluctuation. It is always pleasant to be able to confirm such statements with facts. An extensive orange grower in Putnam County has kept, from the beginning of his grove, an accurate account of the expenditures and receipts to the close of the thirteenth year, ending 1879. The number of trees were 300. They yielded 442,600 oranges, selling for $7,590, as against an expenditure, omitting cost of land, first cost of trees, and interest on the money, of $1,950. This gives receipts over expenditures, $5,640. This is only one instance, but it is as good as many, because it is only one in a very large number. It conclusively demonstrates that orange culture is not at all transitory. Nearly all the obstacles in the path of orange culture have been removed.

The future of the business is still more promising. Florida oranges are conceded to be superior to all others. In point of numbers, compared to the great quantities consumed, they are few; yet by their greater merit they have come to occupy the foremost place in the market. The genial climate and peculiar soil of Florida, together with sufficiently warm sun to mature and concentrate the juices without destroying the lively aromatic flavor of the fruit, impart this quality—a value nowhere else attainable to such an extent. The field they are yet to occupy is practically illimitable. They are yet to possess our own market, the best in the world. This will be the labor of years, and after a great portion of our orange lands have been brought under cultivation. In 1879, there were 257,000,000 of oranges entered at the port of New York alone from foreign countries. Double the number, at least, were entered at all the other ports, making a grand total of 771,000,000 consumed in and lost on the voyage to this country, in addition to our Florida crop. We cannot predict when the domestic will take the place of the foreign product, but it is inevitable in course of time. Our inability to supply the demand is the main obstacle.

That this will be the ultimate result is clear from another cause, independent, or nearly so, of merit. The liability of loss and damage resulting from uncertainties of a sea voyage forms an important factor in the conduct of the foreign fruit-trade, serving to make it extremely hazard-ous—a circumstance against which dealers do not have to contend in the shipment of Florida oranges. We have railroads leading to all the great markets in America, and when the fruit is transported by water, all the facilities are afforded by perfect and commodious steamships.

Orange culture, therefore, may go on indefinitely in Florida, without fear of reaching a general redundance of product. When our own market is occupied those of Europe and elsewhere will be open to us. The grow-ing desire everywhere, also, of people for semi-tropical fruits, which the efforts of producers are trying to satisfy, is unlimited, and, therefore, efforts in orange culture can continue to be put forth until this unlim- ' ited and independent desire is met—a goal which, perhaps, never can be reached.

To persons of foresight and capital, who are looking to the future rather than the present for remunerative returns, Florida presents, in her orange pursuit, the most extended as well as the most inviting field. But aside from the question of profit the culture of oranges pre-sents other practical advantages. It is not only a pleasing but an in-dependent occupation. Its pursuit is no dead level of monotonous ex-ertion, but one that affords scope for the development of an ingenious mind. As a producer, the orange grower is working under conditions of constantly increasing advantages. Young men, sometimes with little or no capital, are starting every year in the business, often away from communities of old and experienced growers, and have succeeded by dint of tact and industry. In point of regular profits; in point of an industrious, frugal, and cheerful occupation; in point of a very general desire to become independent; in point of success and freedom from penury, and in point of repressive and adverse influences in other pur-suits, they have found orange culture, and its practical workings, the most pleasing of occupations. Persons who own groves in Florida are entirely well satisfied, as a rule, with their investments. A bearing grove is worth a great deal of money, and to purchase one would require a large cash outlay. In ten years' time groves are usually in full bear-ing—often in less time—and the inducement to plant one is very great.

Finally, we would say, that the motives that induce men to labor in Florida are the same as in other States—for profit; and if the energy and persistence of the work be proportionate to the constancy and press of the motives, then will they most certainly succeed, and make the aggregate profit of their investment equal, if it does not exceed, that of nearly all other pursuits involving no greater outlay of money. More-over, the occupation of orange-growing has a tendency to make one hopeful for the future. The tilling, too, of the soil immeasurably im-proves the character of the cultivator. Add to this the beauty of the country and climate, and the attractions of country life; the tranquillity* of mind which they promise, and the enjoyments which they really afford; the charm of proprietorship and self-guiding exertion, and the buoyancy of outdoor employment, and we have all the essentials for acquiring health and happiness, as well as independence.

WHAT THE POOR IMMIGRANT MAY DO.

In previous pages we briefly made some remarks as to new-comers. We believe that a plain relation of what may be reasonably assured to the poor as well as rich immigrant will be received as useful information. Florida is no exception to other countries, and the present but repeats the past in the various phases of immigration. The early colonists and colonies in America, the periodical and frequent later immigration to new States and Territories, and from old to new localities, all have had experiences, good, bad, and indifferent, yet we find, after a brief period, that the new countries are filled up with a prosperous and contented population. It is not necessary to review the varied causes of this universal experience; although the local historian may dwell upon them, the new generations of the present look forward and not back. The characteristics of Florida, general and special, we have truthfully noted; other things being equal, the climate, soil, health, cheapness of lands, staple and special productions, easy access and egress by land and water, form of government, low taxes, a small State debt, all present superior advantages, especially for the poor, or those in moderate circumstances, for securing a good home. At the outset, however, the immigrant asks, how shall I at once procure a support for myself and family? Now, premising that the new-comer means to work—intends to stay—he can go to work at once and raise food from the soil. New pine lands, broken up with the grass turned in, will grow good crops of sweet potatoes and cow pease, with but slight cultivation. These crops in, fields inclosed, the grass covered soon becomes rotted, and the soil easily worked. Corn, cane, and cotton may now be planted, as also vegetables; in the same field and with the crops, orange, lemon, and other fruit trees may be planted, where they are to remain, at regular distances apart, both ways. The vineyard may also be put out, as well as smaller fruit, about the premises. The pea-vines, with pease, will afford forage for stock; pease and potatoes for food. Succeeding the pease and potatoes, turnips and onions, beets, cabbage, and similar semi-hardy vegetables may be grown from the late summer to the next late spring months, nearly the year round. The immigrant can easily gather about him hogs, which will range for their own living, potatoes being fed to them in the fall. Poultry are no care for feed or support; game and fish are to be had for the seeking. It will be seen that the food question is easily solved. Year by year his crops are increasing, his comforts added to; he has within himself the accessories of a comfortable home. In the meantime his grove of oranges, lemons, his vines, are growing apace; in a few short years he scents in the early spring the sweet odor of the orange bloom, sees the green fruit gradually increasing in size, and, as autumn months come on, gladdens his eyes with the sight of the golden fruit which now will yield him a substantial return—waited for and won. It has taken less than half a score of years for the piney-woods pioneer to make a new home which yields him ample support and sure increasing income for the future.

WHAT THE RICH IMMIGRANT CAN DO.

To the man of capital, Florida offers a large variety of specialties to employ it surely and profitably, whether as an investment looking to the future for increase, or present employment and quick returns. There are millions of acres of United States, State, and railroad lands, Spanish grants of large areas, and private improved and unimproved lands, which can now be bought cheaply. These comprise timber lands, which are increasing in growth and value every year, also improved lands already cleared and ready to cultivate, now unoccupied by reason of death of owners, or want of means to hire labor and purchase stock. A few thousand dollars judiciously invested in lands would be sure to pay a profit. Manufactories of cotton, and cotton-seed oil mills, starch factories, rice mills, paper mills, tanneries, saw mills, furniture shops, &c., offer good opportunities for present profit.

There are many good openings for mercantile business, purchasing the staples of the country, such as cotton, sugar, sirup, naval stores. Fruit raising on a large scale can be done with assured profit; with means, one can have hundreds of acres in trees, and millions of oranges and lemons to sell or ship. The shrewd real estate dealer can buy and sell at a profit; the money-lender loan at high interest, with ample security. All this has been done, is done, and doing now. If the capitalist would desire to farm on a large scale, no better field than here. There are hundreds of large plantations in Middle Florida, lying contiguous, which can be bought low, and a farm of 100 to 10,000 acres can be made, and planted in cotton, cane, corn, rice, tobacco, and other crops. Labor is plenty and cheap, crops sure and good, always in demand, and fair prices rule.

HOMESTEAD AND OTHER EXEMPTIONS.

One hundred and sixty acres, or one-half acre of land within city or town, owned by the head of a family residing in the State, together with $1,000 of personal property, and the improvements on the real estate, shall be exempted from any forced sale under any process of law; and real estate shall not be alienable without the joint written consent of wife and husband. In addition to the above exemption, there shall be exempted from sale by any legal process, to the head of a family, one thousand dollars in any kind of property which said head of family may select, in cases where the indebtedness was contracted or judgment obtained before the 10th day of May, 1865. Taxes can only be levied for State, county, and municipal purposes. Married women are protected by law in the ownership and control of property separate and apart from the husband.

WHO MAY VOTE.

Every male person twenty-one years of age, who shall be, or shall have declared his intention to become, a citizen of the United States, has resided in the State one year, and in the county six months, may vote in the election district where registered. Bribery, perjury, larceny, wagers on election, fighting a duel or accepting a challenge, disfranchises.

Table showing the mortality and population of the State of Florida for the year ending June 1, 1880, as returned to the Census Bureau at Washington, D. C., and also the deaths from consumption and other pulmonary diseases for the same period.

Counties.	Total mortality.	Deaths from consumption.*	Deaths from pulmonary diseases other than phthisis.	Total population, 1880.
Alachua	68	1	3	18, 597
Baker	7	0	0	2, 312
Bradford	30	3	0	6, 167
Brevard	12	1	0	1, 486
Calhoun	19	0	2	1, 375
Clay	46	5	0	2, 755
Columbia	91	2	8	9, 504
Dade	3	0	0	195
Duval	234	34	11	17, 762
Escambia	107	17	3	12, 217
Franklin	15	1	0	1, 741
Gadsden	191	8	10	11, 588
Hamilton	81	3	1	6, 486
Hernando	0	0	0	4, 254
Hillsborough	40	5	1	5, 888
Holmes	28	1	3	2, 774
Jackson	182	3	22	14, 487
Jefferson	153	10	14	16, 126
La Fayette	11	0	0	2, 000
Leon	282	8	13	20, 325
Levy	78	7	2	5, 776
Liberty	15	1	0	1, 237
Madison	134	5	10	15, 118
Manatee	0	0	0	3, 674
Marion	88	4	2	13, 000
Monroe	62	4	3	10, 927
Nassau	67	6	1	6, 546
Orange	35	6	5	6, 190
Polk	28	1	4	3, 150
Putnam	53	6	3	6, 250
Santa Rosa	19	2	4	6, 701
Sumter	37	0	0	6, 072
Saint John's	58	14	5	4, 595
Suwannee	91	5	4	7, 379
Taylor	45	1	3	2, 280
Volusia	31	0	1	3, 407
Wakulla	11	0	0	2, 750
Walton	31	0	1	4, 270
Washington	31	0	2	3, 238
Total	2, 514	164	141	271, 604

Total deaths in 1,000 of all ages 9
Total deaths from consumption in 1,000 of all ages 0. 10
Total deaths from other pulmonary diseases in 1,000 .... 5. 10

*A large proportion of the deaths from consumption are cases of invalids from other States and countries.

Table showing the population of Florida by race, as returned to the Census Bureau at Washington, D. C., June 8, 1880.

Counties.	White.	Colored.	Total.
Alachua	8,093	10,604	18,697
Baker	1,682	630	2,312
Bradford	4,895	1,272	6,167
Brevard	1,424	62	1,486
Calhoun	979	396	1,375
Clay	2,170	585	2,755
Columbia	4,818	4,776	9,594
Dade	193	2	195
Duval	7,801	9,961	17,762
Escambia	6,988	5,229	12,217
Franklin	1,185	556	1,741
Gadsden	4,072	7,516	11,588
Hamilton	4,334	2,152	6,486
Hernando	3,359	895	4,254
Hillsborough	5,011	877	5,888
Holmes	2,671	103	2,774
Jackson	5,391	9,096	14,487
Jefferson	3,393	12,733	16,126
La Fayette	2,442	158	2,600
Leon	3,440	16,885	20,325
Levy	3,928	1,848	5,776
Liberty	711	526	1,237
Madison	5,656	9,462	15,118
Manatee	3,561	113	3,674
Marion	3,201	9,799	13,000
Monroe	7,668	3,259	10,927
Nassau	2,473	4,073	6,546
Orange	5,494	696	6,190
Polk	3,036	120	3,156
Putnam	3,931	2,319	6,250
Santa Rosa	4,749	1,952	6,701
Sumter	4,993	1,079	6,072
Saint John's	3,242	1,353	4,595
Suwannee	4,166	3,213	7,379
Taylor	2,118	162	2,280
Volusia	2,889	518	3,407
Wakulla	1,460	1,290	2,750
Walton	3,790	480	4,270
Washington	2,380	858	3,238
Total	143,877	127,987	271,864

Population in 1870 187,748

Increase in ten years 84,116

3290——6

Statistics of the State of Florida.

[Compiled by S. U. Hammond, esq., of Fort Gates, from the United States census returns, June 1, 1880, showing the number of acres under cultivation, and the amount of leading productions for the year 1879, together with number and value of live stock, &c.]

Counties.	Land tilled. (Acres.)	Farm values.	Live stock.	Farm productions.	Number of horses and mules.	Number cattle.	Number swine.	Rice. (Lbs.)	Corn. (Bush.)	Oats. (Bush.)	Cotton. (Bales.)	Molasses. (Galls.)	Peaches. (Bush.)	Potatoes. (Bush.)	Garden produce sold.	Honey. (Lbs.)
Alachua	40,771	$701,860	$227,515	$530,433	2,082	8,420	11,245	107,400	258,940	2,040	2,911	107,210	7,450	127,290	$41,355	
Baker	4,877	77,575	53,035	53,650	313	7,432	4,495	29,535	25,507	2,404	206	4,242	7,315	10,720		2,245
Bradford	16,125	231,797	117,813	180,010	875	4,513	9,096	36,000	79,540	4,800		7,700	10,400	34,729		
Brevard	1,950	310,355	171,307	46,840	190	22,358	2,400	222,810	21,012		149			20,230	3,949	3,210
Calhoun	4,306	205,775	42,637	45,972	248	5,702	2,974	21,570	12,012	2,393	86		1,337	22,370		12,343
Clay	3,871	95,550	45,050	63,325	326	2,651	4,430		70,751	4,184	120			14,540		
Columbia	33,006	371,290	190,217	317,825	1,515	8,762	9,797	65,705				7,361	10,300	41,740		1,230
Dade (no returns)																
Duval	4,200	811,745	166,031	43,125	423	3,738	3,254	16,775	9,802		288	7,393	3,805	25,875	13,541	3,085
Escambia	1,963	112,725	23,356	31,000	507	4,309	2,309						6,000	15,305		10,000
Franklin		20,750	13,305	10,580	18	1,869	1,318		761					19,270		2,400
Gadsden	76,997	707,514	106,696	459,710	2,461	14,829	587	41,400	133,325	27,150	5,372	13,630	1,320	75,845	1,520	24,494
Hamilton	48,694	453,335	157,451	190,690	857	10,473	10,493	130,516	110,988	21,303	1,867	53,960	2,340	31,335		9,490
Hernando	12,724	378,000	213,600	207,100	1,080	24,960	28,200	90,000	63,400	20,400	867	26,000		120,000		12,370
Hillsborough	11,447	1,046,265			940	17,210	9,565	11,000	55,480			9,321	4,200	68,287		30,400
Holmes	7,040	40,210	56,840	44,240	675	940	9,000	108,000	57,060	5,000	365	35,200		31,600		2,400
Jackson	128,405	1,173,400	409,276	1,908,481	3,259	12,996	12,467	72,520	407,722	65,412	320	55,140		105,675		2,135
Jefferson	89,830	2,280,070	201,657	736,618	2,206	12,417	12,512	8,400	359,704		11,565	67,396		97,444		
La Fayette	7,782	92,630	80,143	50,220	2,088	6,043	4,778				8,728			11,833		3,780
Leon	104,047	863,360	196,656	663,190	3,031	7,486	12,054	5,545	320,684	29,986	9,072	63,988	2,865	114,197	10,900	
Levy	15,523	268,360	146,315	55,779	900	5,330	5,144		45,386	16,840	440	31,240		48,200		
Liberty	3,863	67,375	44,012	55,600	163	13,427	3,125	24,825	16,285		440	110,901		13,272		6,000
Madison	61,010	996,450	211,900	681,695	2,994	17,402	17,028	10,000	343,050	48,040	6,773	103,611		107,940		14,400
Manatee	3,300	384,000	912,000	177,600	170	190,650	8,400	18,300						85,460		2,025
Marion	49,794	1,113,000	44,012	292,588	2,213	5,700	10,274		187,255	4,400	2,870	33,876	4,907	79,215		
Monroe	906	111,000	12,500	19,290	170	7,516	4,108					12,000		12,000		
Nassau	4,562	136,850	62,182	43,898	303	12,143	3,041	5,700	24,400	2,573	292	10,408		22,212		503
Orange	13,166	3,381,410	294,330	90,025	1,384	7,684	6,531	29,000	30,430	1,116	71	21,210		75,745	2,737	158
Polk	8,160	400,000	56,000	80,000	640	3,012	6,400	4,337	39,380		200	4,860		66,400		7,200
Putnam	13,718	2,151,507	40,980	146,759	567	822	4,589	205,140	27,271	5,133	1,856	3,650		29,091	17,988	5,695
Santa Rosa	2,030	50,500	24,380	34,290	238	6,350	2,609		10,596	900		11,235	8,947	16,240		22,325
Saint John's	2,880	717,100	67,985	86,308	450		3,833	2,300	13,927	445	8	9,550	240	29,876		
Sumter	10,412	474,024	79,500	26,542	1,022		7,632		45,330	6,870	300			23,940		800

HARBOR AND CITY OF PENSACOLA.

The Gulf of Mexico is the natural basin for a larger extent of country than any similar sheet of water on the globe, and the finest bay and harbor on its coast is that of Pensacola, on which are located Forts Pickens and Barrancas and the Warrenton navy-yard.

The city of Pensacola, located at the head of the bay, is a most beautiful place, and deserves special mention as the most attractive feature of West Florida. The following description is derived from an official publication of the Commissioner of Immigration :

The city of Pensacola has natural advantages which destine it to become, by rapid strides, the Chicago of the South. It is situated on the north coast of the Gulf of Mexico, in latitude 30° 28' north and longitude 87° 22' west of Greenwich, only 10 miles from the open sea. Its thoroughly land-locked harbor covers an area of over 200 square miles, being about 30 miles long and from 5 to 8 miles in width, having unsurpassed anchorage, and a depth of from 30 to 35 feet. The entrance to the harbor is about half a mile wide, with an average depth on the bar of *twenty-four feet.* The same depth is readily secured at the wharfage line in the city. A laden ship of the largest tonnage can approach the city at any time in the year, or leaving its wharf can be in the open sea in an hour and a half.

As a place of residence, Pensacola is attractive by a healthy and genial climate. It has an abundance of excellent pure water, and the regularly changing land and sea breezes make it a pleasant abode at all seasons. Its gardens afford flowers and fruit winter and summer. Most tropical plants grow there, and require but little protection from the cold in winter, whilst all kinds of cereals and northern fruit yield early and abundant crops. The soil of the immediate vicinity is sandy and the drainage perfect.

The lands of the neighboring country are of the character known as swamp, hammock, and pine. The swamp lands are those lying on the Escambia and Perdido Rivers and their tributaries, and are remarkable for their exhaustless fertility, those brought under cultivation yielding heavy crops of corn, cotton, rice, and sugar-cane. The great body of these lands is covered with oak and cypress forests, ready to the hand of the great ship-building interests, which the progress of commerce will speedily foster.

The hammock lands are intermediate between the swamp and pine tracts. They afford the healthiest localities for settlements, and are easily cultivated, yielding abundant returns for the labor bestowed on them.

The pine lands have an exhaustless wealth of the best timber, whilst the herbage of the forest affords permanent pasturage for cattle, which require no shelter and very little care.

All these classes of lands are readily reclaimed for agricultural purposes, whilst the forest will for a century to come supply the growing export trade in timber.

The commerce of Pensacola has hitherto been limited to the export of Florida timber brought down on the rivers and creeks emptying into its magnificent bay. Want of communication has been an impediment to its growth, but the completion, in the winter of 1870, of the Pensacola and Louisville Railroad to its junction with the Mobile and Montgomery Railroad, establishing *a connecting link with the entire railroad system North and West,* must speedily lift Pensacola to the dignity of a first-class commercial city. By this link in the great chain of railroads, Pensacola is brought as near to Chicago as is New York.

The best customers and consumers of the great grain and produce growing West live upon the shores of the Gulf, in the West India Islands, in Central and South America. The Pensacola route of transportation shortens the exchange of commodities between these markets and the teeming West by thousands of miles and by many days, thus effecting a double economy of time and cost of transportation.

A glance at the map will readily demonstrate the fact, so little known heretofore, that the distance from Chicago to Pensacola is only about 900 miles. It will also show that from Pensacola the distance to Tampico is 900 miles; to Havana, 620 miles; to Matamoras, 800 miles; to Vera Cruz, 950 miles; to Hansacula, 950 miles. The last-named place is the eastern port of the Isthmus of Tehuantepec.

No vessel has ever been driven ashore in any storm in the harbor of Pensacola, and no gale has produced a freshet. The rivers emptying into the bay are clear and free from alluvial deposit, and the depth of water on the anchorage ground, as well as on the bar, remains unaltered since the earliest Spanish surveys.

With the railway connection recently established and daily expanding, this magnificent port becomes the most suitable outlet f r the commerce of the entire Mississippi Valley. It has this great advantage over New Orleans, that it is close to the Gulf, and not obstructed in its commerce by a shifting and treacherous bar, causing costly delays to shipping, and often upsetting the fairest calculation for commercial profits. The vast expenditure in towage up and down the river to which the New Orleans shipping is subjected in reaching and leaving that inland port is saved in Pensacola, and it is easy demonstrable that shippers in New Orleans can ship their cargoes more cheaply from the port of Pensacola than from their own levee. Still greater will be this economy when the canals now proposed and under survey shall connect the Mississippi with Mobile Bay, Perdido Bay, and Pensacola Bay, permitting steamers to bring their upland cargoes directly to Pensacola, and lay them alongside the sea-going vessels.

The splendid water-front of the city admits of running railway freight directly down on the wharves, and to load vessels immediately from the cars. The elevated bluffs on this water-front affords facilities for coal depots, from which vessels can be supplied through shutes, thus saving greatly in expense of handling.

Having thus briefly alluded to the physical features of the port, we will now examine the advantages of its relative position to other ports.

Taking Chicago as the initial or starting point, as being equally distant from New York and Pensacola, railroad trains destined to each of the cities would arrive at their destination within the same time. The one arriving at New York would have traveled over 900 miles, and would then be as *far north* as when it started from Chicago, whereas the one arriving at Pensacola would have gone directly south 900 miles, thus saving that number of miles between the initial point (Chicago) and any other point south of Pensacola. This distance, to be balanced by transit to and from New York, is equal to a gain of six days in favor of Pensacola.

Take now the return cargoes, one *via* New York and the other *via* Pensacola, say coffee, &c., from Havana, distant from Pensacola 620 miles. The one by way of Pensacola would have reached its ultimate destination, and have been distributed, before the other could possibly have reached New York. These remarks apply with equal force to all the cities and towns lying along and in connection with this great national artery of intercommunication, trade, and commerce.

The Pensacola and Louisville Railroad line and its connections, unlike those leading to the Atlantic ports, *bisect* the parallels of latitude of the United States; hence it must collect and transmit the productions of these different latitudes, consisting of wheat, flour, corn, pork, bacon, lard, cheese, bagging, rope, iron, lime, coal, and a great variety of industrial products, such as furniture, clothing, machinery, implements, &c., concentrating them all by one line at one single point of shipment, and giving that point the same advantages to be offered to the shipping interests of the world that are now afforded at the said Atlantic ports through a hundred different channels at a vastly increased expense, both in time and money, and enabling ships desiring freights to any part of the world to make such selections as their interests or exigencies may require.

The commerce of the world will hereafter be carried on through the agency of steam, and will expand in the use of that agent just in the ratio in which fuel (coal) can be easily and cheaply obtained for that purpose. The coal-beds of Alabama, estimated to cover between 4,000 and 5,000 square miles of area, are intersected by this line of road, and, from their contiguity to Pensacola, must become the great source of supply for the steam marine and coaling stations of all points south of Pensacola. The coal now used for this purpose is principally brought from Great Britain, a distance of 3,000 miles. From the Alabama coal-beds to Havana (which can be thus supplied) the distance is about 810 miles, and coal from these mines can be placed on shipboard at Pensacola at $4.75 per ton ; the sea transportation is but 620 miles. These facts and figures guarantee that the day is not far distant when Pensacola must become the great coal depot of the Gulf, making annual shipment of millions of tons of this article, developing the resources and wealth of the country, and expanding into one of the first cities in the world.

The rapid development of the iron mines of Alabama, whose natural outlet to the markets of the world is the port of Pensacola, will not only contribute a considerable quota to the commerce of this port, but will, in connection with the Florida forests, furnish superior material for ship-building, which at no distant day must rival in extent the similar industry of Northern ports, the proximity and cheapness of all material required, giving builders in this locality peculiar advantages.

FACILITIES FOR TRANSPORTATION.

No State in the Union has so extended a sea-coast as Florida, and none contains a larger extent of inland navigable water; nor is there any State which enjoys greater facilities for permanent, reliable, and cheap communication with the commercial marts of the world and the interior cities of the North and West. Ocean steamers, with the most ample accommodations for passengers and the most extended appointments for freight, ply regularly between New York, Boston, Philadelphia, Baltimore, Charleston, and Savannah, and the Florida Atlantic ports. At Fernandina these lines connect with the Gulf and West India Transit Railway, which, at Hart's Road, connects with the Jacksonville and Fernandina Railway ; at Callahan intersects the Savannah and Jacksonville Railway ; at Baldwin with the Florida Central Railway ; at Waldo connects with the Peninsular Railway to Ocala, and with the Santa Fé Canal to Santa Fé Lake; at Gainesville with the Florida Southern Railway to Palatka and Ocala ; and at Cedar Keys with lines of steamers to Tampa, Key West, Havana, New Orleans, and all the Gulf ports.

At Jacksonville connections are made with the numerous steamers on the Saint John's and Oclawaha Rivers, which connect at Tocoi with the Saint John's Railway to Saint Augustine; at Palatka with the Florida Southern Railway to Gainesville and Ocala; at Astor with the Saint John's and Lake Eustis Railway; at Sanford with the South Florida Railway to Lake Apopka and the Kissimmee River ; at Salt Lake with the Saint John's and Indian River Tramway to Titusville; and at Lake Poinsett with lines of stages to the Indian River at Rock Ledge.

At Jacksonville connections are also made with the Fernandina and Jacksonville Railway ; with the Savannah and Florida Railway, which intersects the Gulf and West India Transit Railway at Callahan; with the Florida Central Railway. which intersects the Gulf and West India Transit at Baldwin; connects at Lake City with the Jacksonville, Pensacola and Mobile Railway, which, at Ellaville, intersects the Suwannee River, which is navigable for steamers to Cedar Keys; at Live Oak connects with the Savannah, Florida and Western Railway; at Tallahassee with the Railway at the Gulf at Saint Mark's; and at Chattahoochee with the Atlantic and Western Railway, now in process of construction to Mobile, and with lines of steamers to Apalachicola, Eufaula, Ala., and Columbus, Ga. All rail routes, with close connections and through parlor and sleeping cars for passengers and fast freight lines, with ventilated cars for fruit and vegetables, connect Florida with Montgomery, Atlanta, Louisville, Cincinnati, Saint Louis, and Chicago, in the West; Savannah, Charleston, Richmond, Washington, Baltimore, Philadelphia, New York, and Boston, in the North, thus affording the largest facilities for rapid transit with the numerous competing lines, and the ocean steamers prevent exorbitant charges.

The completion of the lines of railway now under construction will render all portions of the State immediately accessible. The State authorities have granted about 16,000,000 of acres of her swamp lands in aid of works of internal improvements, thus giving a powerful impetus to these enterprises, while Northern capital is pushing them to rapid completion.

No less than five lines of railway are now in progress in East Florida, running south, with the view of reaching Tampa Bay, Charlotte Harbor, Indian River, and Key West.

1st. The Tropical Railway, an extension of the Peninsular, is progressing rapidly from Ocala southward, with its iron purchased, and more than a thousand hands at work.

2d. The Florida Southern, from Palatka to Gainesville, is pushing south also from Ocala to the eastward of the Peninsula.

3d. The South Florida, from Sanford, is extending southward still farther to the east.

4th. The Live Oak, Rowlands Bluff and Charlotte Harbor Railway is backed by a powerful company.

5th. The Jacksonville, Saint Augustine and Halifax River Railway is under construction, from Jacksonville to Saint Augustine, with a view to an ultimate extension down the coast to the Indian River.

The Atlantic and Western Railway, an extension of the Jacksonville, Pensacola and Mobile Railway, from Chattahoochee to Pensacola, is under contract and being rapidly constructed, with a view to immediate completion.

A large portion of the route of the Jacksonville, Tampa and Key West Railway has been surveyed, and no doubt exists of the early commencement of that important enterprise.

Those in process of construction, and others projected, with the system of canals connecting the lower Saint John's with North Halifax and Indian Rivers, and the Upper Saint John's with the Kissimmee, Lake Okeechobee, and Caloosahatchee, when all completed, will give Florida the most complete system of internal communication of any Southern State, and render her one of the wealthiest and most desirable for residence and cultivation.

FISHERIES.

The extent and wealth of the fisheries of Florida are, beyond comparison, greater than any other State of the Union. Her twelve hundred miles of sea-coast from Fernandina to Key West, and round to Cedar Keys and Pensacola, with the extensive bays and harbors, abound in turtle, oysters, and sponge, while the waters teem with fish in the greatest variety.

It has been remarked by an experienced observer that a fee-simple to three miles wide of her coast line of waters was more valuable than the same amount of land adjacent. The annual exportation of oysters. fish, and sponge amounts to hundreds of thousands of dollars. Hundreds of families in Florida, Georgia, and Alabama annually resort to the sea-coast and gather a supply of fish, with which they return home.

State legislation has, as yet, furnished no adequate protection for these fisheries, and they are annually used by fishermen from other States, " without money and without price," and the State derives no revenue.

The inland waters. too—the rivers, creeks, innumerable lakes and ponds—all abound in the varieties of fresh-water fish, which are gathered without let or hinderance for food and enriching the soil, while the supply seems inexhaustible. With proper care and protection against waste and destruction, the supply would last for generations before it became necessary to resort to artificial cultivation.

STOCK GROWING.

Cattle raising has long been one of the most lucrative branches of business in Florida. In large portions of the State, notably in the southern, the growing of crops has been neglected for the all-absorbing business of cattle raising, which is attended with no expense, save the personal supervision of the herds, and in gathering the stock for marking and for market. The cattle range on the public domain through the year, and the plains, savannas, and swamps of South Florida have afforded pasturage for innumerable herds, from which, during the civil war, the Southern army drew large supplies, and from which the markets of Key West and Cuba derive their present supply. It is not uncommon to find men owning thousands of head of cattle without the proprietorship of lands, and many of the herdsmen count their stock by tens of thousands. The profits are fabulous, as the cost of. keeping the cattle is only the expenditure for herding and marking, no food being required other than the natural supply. The annual burning of the grass, which has long prevailed in the winter, has seriously impoverished the soil and reduced the quantity of grass, so that the native stock of later grass has become dwarfed and inferior. As population increases, and the lands become absorbed by settlers, the stock range becomes more restricted, improved methods are being adopted, and improved stock introduced. No State offers greater inducements for stock growing, either upon the wild method heretofore existing, or upon the system of thorough cultivation and high breeding. The facilities for cultivating green forage crops during the winter, when the grasses become tough and innutritious, afford great advantages over those sections of the country where cattle must be housed and fed for from one-third to one-half the year.

The raising of sheep and growing of wool is also a most profitable industry, as sheep thrive through the year on the natural pasturage, and require no care except herding and protection from vagrant dogs, of which there is too great a number. With no legal protection against these depredators the business, when it has been prosecuted as a reliance, has yielded from 33 to 90 per cent. per annum profits. It can be seen at a glance, that with an annual increase very largely greater than in the cold States, and no expense for feed, and entire exemption from the cold northers which sweep off whole flocks in the West, the profits must be very great and the industry so remunerative as to induce a rapid extension of the business.

Hogs are raised more cheaply and easily here than in the colder climates. They do well "on the range," as in the winter there is a large supply of acorns, and in the swamps and hammocks of roots and native products. There is no reason why pork should not be raised in sufficient quantity to supply the home market at least, although the want of frost, or freezing weather, is not conducive to profitable packing for export.

Bee-keeping is rapidly engaging attention, and will soon become a source of large State wealth and individual revenue.

THE TREES OF FLORIDA.

Dr. A. W. Chapman, of Apalachicola, author of the Flora of the Southern States, in 1875 made a journey along the western coast, for the purpose of obtaining specimens of trees for the Centennial Exhibition. He made a thorough exploration from Anclote Keys to Cape Sable, and ascended the Caloosahatchee. The following is the list of trees obtained, as given in his report:

The number collected exceeds your estimate by ten species, and falls short of my own by the same number. I believe I obtained all the native trees known down there, except *Simaruba*, and perhaps *Calyptranthus*, if it is a tree. I found several trees which I supposed to be shrubs, and *vice versa*. I made a thorough exploration of the whole western coast, from Anclote Keys to Cape Sable, wherever we could find smooth water for safe anchorage. At Charlotte Harbor I diverged from the coast and ascended

Caloosahatchee River, in order to get such woods as do not grow in the influence of salt air. This was really the most interesting part of the route. Gigantic Acrostichums, 10 feet high, covering acres, Epiphtyes loading the trees, and the entire vegetation tropical. A peculiarity of these tropical trees is that for miles they occur to you as mere shrubs, when at some other locality you find them lofty trees. I was much disappointed in the size of most of the forest growth in that region. On the Keys you can scarcely anywhere find a large (or rather a tall) tree. Some of these were large enough at the base, but we generally found such hollow, and of some we never did find a sound one. You will be disappointed, as I was, to find the growth so small. I do not remember to have seen a tree, during the trip, 2 feet in diameter, with the exception of the Live Oak, and on the Keys none of them get to be more than 30 or 40 feet high. The Mahogany is not found in Florida, and should be erased from the Flora. My authority for introducing it was a pod picked up on the beach by Dr. Leitner.

Hibiscus tiliaceus was not seen by me, and I think Dr. Blodgett must have got it from cultivation. In Jamaica it is a shrub 12 to 15 feet high. *Terminalia* is not a native, and is, I believe, local along the Saint John's or near Saint Augustine. The others mentioned I did not meet with on any of the Keys I visited. Whether they become trees I cannot say, for I forget the sources of my information regarding them when writing my book. It was of course impossible to visit all the hundreds of Keys along the reefs, and it is probable that these omissions may occur on more westwardly ones than those I visted.

The following is a modified arrangement of Dr. Chapman's list of Florida trees:

Anona (Custard Apple).—No flowers or fruit; 15 to 20 feet high. The fruit is said to be egg-shaped, 1¼ inches in diameter, and eatable when fully ripe.

Capparis Jamacencis (Caper-tree).—A low tree.

Canella alba.

Guaiacum sanctum (Lignum Vitæ).—Only found, if I am rightly informed, on the Lignum Vitæ Keys, and quite rare there.

Xanthoxylum Pterota, 12 to 20 feet high.

Bursera gummifera (Gumbo Limbo, Gummer Limmer).—The largest of South Florida trees, abounding in gum.

Amyris Floridiana (Torch-wood).—Mostly shrubby.

Xymenia Americana (Hog Plum).—2 to 20 feet high.

Schaefferia frutiscens (Crab-wood).—A small tree.

Sapindus (White-wood).—This is scarcely the tree of the Southern States and of my Flora; I suspect it is *S. saponaria.*

Hypelate paniculata (Maderia-wood).—This wood is very like Mahogany, and is highly valued. It is not abundant, and was only found on Metacumba Keys.

Rhus metopium, 20 to 30 feet high. It is very poisonous, and we all got peppered by it.

Piscidia erythrina (Dog-wood).—A rather large tree.

Pithecolobium unguis-cati.—Rarely a small tree.

Rhizophora mangle (Red Mangrove).—Commonly a low spreading tree. On the Thousand Islands it attains its largest size—40 to 60 feet. All the low Keys are formed by this tree.

Conocarpus erecta (White Buttonwood.)—It comprises almost the only fuel used in Southern Florida, and extends north as far as Anclote Keys.

Luguncularia racemosa (Black Buttonwood).—A small tree everywhere, or a mere shrub, except among the Thousand Islands and north of Cape Sable, where it forms a large tree.

Eugenia buxifolia (Iron-wood).—25 to 30 feet high.

Eugenia Monticola.—South Florida; about 30 feet high.

Eugenia (——).

Eugenia (———), uear *dichotoma*, but probably distinct. This was only seen at Caximbus Bay, and was called "naked-wood."
Eugenia (Stopper-wood).—A small tree, in fruit.
Guettards Blodgetti.—Mostly a shrub.
Randia clusiæfolia (Seven-years' Apple).—With flowers and fruit.
Sideroxylon padillum (Mastic).—A middle-sized tree.
Sideroxylon padillum, var. *sphærocarpum.*—A small tree.
Sideroxylon.—A large tree.
Chrysophyllum microophyllum.—6 to 20 feet high.
Mimusops Sieberi.—A large tree. We found the trunk invariably hollow.
Bumelia parrifolia.—A shrub or small tree.
Jaquinia armillaris.—A rather small tree with most curiously grained wood.
Myrsine Floridana.—Mostly a shrub; rarely a small tree.
Ardisia Pickeringii.—Mostly a shrub, but on the Keys a small tree.
Citharexylon villosum (Fiddle-wood).—Rarely a small tree.
Avicenna oblongifolia (Black Mangrove).—Only a tree among the Thousand Islands.
Avicennia tomentosa (Black Mangrove).—At Cedar Keys only.
Pisonia obtusata.—With male flowers.
Coccoloba Floridana.—20 to 30 feet high. In fruit.
Coccoloba ubifera (Sea-side Grape).—In fruit.
Persea Catesbaei.—No flowers or fruit.
Drypetes crocea.—A small tree.
Ficus aurea (Wild Fig).—A large tree full of milky juice. It is also called gum-tree, and the juice forms a kind of India rubber.
Ficus.—Perhaps the same as the preceding.
——— (Silver Palmetto or Silver Cabbage Tree).—The berries are white, but in the absence of flowers the genus is doubtful. It attains a height of 30 to 40 feet. It occurs first at Cape Romano, and is found sparingly on the mainland southward. It is more common on the Keys, but I never heard of it before.
Yucca aloifolia. I found this from Manatee southward, 15 to 25 feet high.
Pinus clausa, N. sp.—At Apalachicola. Dr. Engelmann is doubtful. Perhaps it may be a variety of *P. inops.*

ORANGE STATISTICS OF FLORIDA.

Judge A. A. Knight, the supervisor of the Tenth Census for the State of Florida, was, in addition to his other duties, intrusted with the task of ascertaining the number of bearing orange trees in the State, and their production for the year 1881. He has kindly permitted me to take the following figures from the papers in his office. Twenty-five thousand circulars were issued, which elicited about 70 per cent. of replies. Twenty-five of the thirty-nine counties in the State are embraced in the report, fourteen failing to return replies. The deficiency in this direction is conterbalanced by the supplement which closes the report· It is expected that the yield of fruit will very rapidly increase from this time forward, both from the large number of new groves coming into bearing with each succeeding year, and the increased productiveness of the older trees.

REPORT.

Counties.	Number of bearing trees.	Yield in 1881.	Value.
		Number.	
Alachua	13, 111	2, 250, 000	$33, 750 00
Baker	21	9, 450	141 75
Bradford	3, 377	338, 850	4, 815 50
Brevard	10, 884	1, 250, 000	18, 750 00
Calhoun	841	282, 400	4, 170 50
Clay	738	165, 700	2, 522 25
Columbia	436	157, 850	2, 741 00
Dade	500	500, 000	7, 500 00
Duval	10, 131	3, 000, 000	45, 000 00
Escambia	No report		
Franklin	do		
Gadsden	do		
Hamilton	do		
Hernando	7, 685	2, 500, 000	37, 500 00
Hillsborough	18, 683	4, 409, 150	45, 410 25
Holmes	No report		
Jackson	do		
Jefferson	do		
La Fayette	1, 157	43, 800	662 00
Leon	No report		
Levy	1, 460	500, 000	7, 500 00
Liberty	No report		
Madison	594	512, 900	7, 685 00
Manatee	17, 291	2, 000, 000	30, 000 00
Marion	46, 195	6, 000, 000	90, 000 00
Monroe	500	500, 000	7, 500 00
Nassau	No report		
Orange	29, 049	4, 000, 000	30, 000 00
Polk	2, 283	1, 500, 000	22, 500 00
Putnam	64, 170	7, 120, 631	108, 414 80
Santa Rosa	No report		
Saint John's	12, 006	2, 000, 000	30, 000 00
Sumter	13, 029	2, 250, 000	33, 750 00
Suwannee	157	120, 700	2, 060 00
Taylor	1, 846	255, 200	2, 747 50
Volusia	24, 638	4, 000, 000	60, 000 00
Wakulla	No report		
Walton	do		
Washington	do		
Supplement	11, 536	451, 225	7, 056 10
Total	292, 324	46, 097, 856	672, 176 65

Respectfully submitted.

GEORGE B. CARSE.

FLORIDA: ITS SOIL, RESOURCES, AND MEANS OF TRANSPORTATION.

For reliable information in regard to all points of agricultural interest I must confine myself in this report to the counties of Nassau, Duval, Clay, Saint John's, Volusia, Orange, Marion, Brevard, and Putnam.

NASSAU COUNTY.

This county contains an area of 600 square miles, and has but about 4,000 acres under cultivation owing to the fact that much of the land is "pine barrens" and "flat-wood" country, of but little value to agriculturists. Scattered along the river is some land which produces grazing for cattle, a limited amount of long-staple cotton, corn, potatoes, and peaches. Here fertilizers are necessary, and winter vegetables are liable to be nipped by the frost; still, the annual yield of farm products reaches the value of $50,000, consisting chiefly of cattle, swine, poultry, corn, potatoes, and peaches.

This county is well supplied with railroads, and may have a prosperous future, should its people, the great majority of whom were once slaves, ever attain to that state of intelligence indispensable to agricultural success. Fernandina, the county seat, has a fine harbor and quite a lumber trade.

This county cannot be recommended for the raising of oranges or of the semi-tropical fruits, but may be reasonably successful with the ordinary farm crops.

DUVAL COUNTY.

With an area of 860 square miles this county has only about 5,000 acres of land under cultivation. While the greater part of the soil is light and sandy, there are some tracts of rich "hammock" which may be utilized for the cultivation of rice, corn, potatoes, cane, and cotton; but it is to much exposed to the cold northwest winds for the successful raising of oranges or of the semi-tropical fruits, with the exception of that portion east of the Saint John's River. Here, as elsewhere in this State, the eastern shore of the Saint John's is much better protected from the cold than the western banks, owing to the fact that the cold northwest winds are considerably warmed in passing over this wide expanse of water. Although the county is much frequented by invalids, the climate is not conducive to the health of those suffering from pulmonary or bronchial diseases, as the changes in temperature are very sudden and severe, ice being frequently formed near the mouth, of the river, immediately followed by very hot weather. The annual farm productions are valued at about $50,000, consisting principally of cattle, milk, butter, swine, poultry, rice, cotton, potatoes, and molasses. The means of transportation in this county are ample, both by rail and by water, the rates here, as elsewhere in this State, averaging about five cents per mile by rail, and three cents per mile by steamer. Jacksonville, with its numerous hotels, furnishes a good market for garden farmers. Land in the city commands very high prices, building lots being frequently sold at the rate of several thousands of dollars per acre. Skilled farmers, with the aid of fertilizers, can do well in some parts of this county.

CLAY COUNTY.

With an area of 425 square miles, Clay County has about 4,000 acres of land under cultivation, and raises farm productions to the value of about $64,000 yearly, including some cotton and sugar. There are several fine lakes in this county, containing many excellent food-fish; but lands must be selected here with great care to avoid malarious districts, and it should not be recommended for raising fruit of the Citrus family, though many trees are being set out near Green Cove Springs, the county seat, which may succeed in sheltered places protected by heavily-timbered pine lands. There is considerable yellow-pine timber in this region, with some rich hammock lands, well adapted to the production of the ordinary farm crops. No railroad is found in the county, and it has but a small population, with no large towns. There are several hotels at the county seat, but those afflicted with pulmonary or bronchial difficulties should go farther back among the pines, away from the chilling, damp night air of the Saint John's.

SAINT JOHN'S COUNTY.

This county contains 970 square miles, and has but about 3,000 acres under cultivation, owing to the fact that it is mostly a " flat pine-woods and palmetto-scrub country," with but little rich hammock land. Its location, however, between the Saint John's River and the Atlantic renders it more exempt from frost and better adapted to fruit culture than more interior counties in the same latitude. Colonel Hart's famous orange groves are located in this county, opposite Palatka, and demonstrate the fact that with skillful cultivation and the aid of fertilizers excellent oranges can be raised here. The farm productions amount to over $90,000 per annum, including considerable rice, sugar, cotton, potatoes, and corn. Here skilled labor, backed by energy and sufficient ready money to tide the settler over for a few years, will in due time reap its' reward; but the impecunious and the ignorant will find no bonanza in Florida. This county has but one railroad, that connecting Saint Augustine with the Saint John's at Tocoi, fourteen miles in length, over which the fare is $2.

MARION COUNTY.

Comprising an area of 1,000 square miles, and having nearly 50,000 acres under cultivation, this county contains some of the richest and most elevated land in in the State. The high hammock lands are exten·sive and very productive, but much better adapted to the raising of vegetables than to the development of healthy men, and much care must be taken in locating dwellings, or, as in the West, fever and ague will take off all the profits for the benefit of the doctor. Although there are some flourishing groves in sheltered places near Ocala, oranges there are liable to suffer from frosts, and the extreme eastern and southern portions of the county are better suited to the culture of the Citrus family of fruits. Farm productions are raised in this county to the value of $300,000 yearly, including principally corn, potatoes, rice, molasses, and poultry. Lands near transportation are held at high figures, and there is but little government land to be obtained.

Two railroads run through this county, and the means of transportation will soon be exceptionally good *via* these roads and the Oeklawaha River, which is navigated by two lines of steamers from Silver Springs to and over the Saint John's. There are some beautiful lakes in this county, the eastern shores of which are well adapted to orange culture, being thus protected from the cold northwest winds. Considerable game is found in the heavy-timbered tracts, and food-fish abound in the lakes and rivers. The Florida Southern Railroad, owned by Boston capitalists, is developing an extensive and profitable lumber trade along its line, and many of the settlers build houses, barns, and fences with the beautiful yellow pine, green from the saw-mill. Industrious settlers who can afford to wait until the acidity of the soil is removed by culti·vation will succeed here, provided they can secure good land at reason-able prices. No injurious acids are found in the hard-wood lands, but unless great care is taken the farmer will suffer from chills during the first year's cultivation.

BREVARD COUNTY.

Containing an area of 4,000 square miles, this county has but 2,000 acres under cultivation. Much of this county is composed of " flat lands," often overflowed and not easily drained, with a stiff-clay subsoil, through which the tap-root of the orange tree cannot penetrate to water; consequently the Citrus family of fruits will not thrive in the greater part of this region. Along the banks of the Indian River, however, excellent oranges and semi-tropical fruits are raised ; but the land in this county adapted to such culture is limited in extent, and is held at high prices. Oysters, fish, and game abound in some localities, and much good grazing land is found ; but malaria must be carefully guarded against, and there is but comparatively little land in the county adapted to the wants of people from the North for a permanent residence. Yearly

farm productions are valued at about $47,000, including principally
cattle, swine, rice, and potatoes. The South Florida Railroad extends
40 miles from Sanford, on Lake Monro, Orange County, to Lake Kissimee, Brevard County. Titusville, the county seat, is reached by a
long steamboat ride from Lake Monro to Rock Ledge, and thence by
carriage. Drainage may open large tracts of rich lands to the settler;
but until then this region is more attractive to the sportsman than to
the agriculturist.

VOLUSIA COUNTY.

This county contains 1,800 square miles, and has but 4,000 acres
under cultivation. The eastern portion along the Halifax River comprises some rich hammock lands, which in years past have produced
great quantities of sugar, and with good cultivation may again yield
much cane. West of this belt is a vast prairie, interspersed with pine
and cabbage palmetto, affording excellent pasturage to large herds of
cattle during the entire year. On the extreme west, extending from the
northern end of the county south about 30 miles, is a rolling pine country,
on which are hundreds of young orange groves; from this southward
is a rolling pine scrub until the Saint John's is reached, with its varying banks of hammock and savanna.

With much care to avoid miasmatic swamps, and the chills which
arise from newly-plowed rich hammock lands, desirable farms may be
found in this region, which, when cleared and dispossessed of their acidity
by cultivation, will produce semi-tropical fruits and vegetables in abundance. At present the yearly farm productions are valued at $60,000,
including principally cattle, poultry, cotton, molasses, and potatoes.
The present yield of lumber is one and one-half millions of feet per annum, which can be increased almost indefinitely.

Transportation is confined to steamers on the Saint John's River, but
several railroads are chartered. Enterprise, the county seat, has demonstrated the important fact that oranges and vegetables can be raised at
a profit by skilled laborers, with the aid of fertilizers, the most popular
of which is that made by George B. Forrester, of New York.

ORANGE COUNTY.

Orange County, comprising 2,300 square miles, and having 14,000
acres under cultivation, is generall rolling pine land, interspersed with
large lakes, rich hammocks, and comparatively worthless flat pine lands,
all more or less heavily timbered. Stock-raising has been the predominant
industry until recently, with cotton, corn, and cane; but now fruit culture is absorbing general attention, and the orange, lemon, lime, guava,
pine-apple, and banana are cultivated to considerable extent. The South
Florida Railroad runs from Sanford southerly through the county, which,
with a short road from Astor, on the Saint John's, to Lake Eustis, together with the numerous Saint John's steamers, furnish exceptionally

good facilities for transportation. The more desirable lands near the towns are held at high prices; but still, by using much care to avoid unhealthful localities industrious and skillful agriculturists may obtain good homes. Successful ice manufactories have been established, and by using cistern water many Northerners live here all the year in tolerable comfort. The yearly farm productions, consisting chiefly of cattle, poultry, rice, corn, potatoes, molasses, and honey, are valued at $100,000, and many fine orange and lemon groves may be seen about Sanford, Orlando, Bellaire, and Altamont.

PUTNAM COUNTY.

Comprising over 800 square miles, with 1,500 acres under cultivation, this is in many respects the most desirable and flourishing county in the State. The numerous lines of steamers on the Saint Johns' together with the South Florida Railroad, afford facilities for transportation unexcelled in the State. Many beautiful lakes are found in this region, fully stocked with excellent food fish, and game in many parts is abundant. The portion of the county lying east of the Saint John's is appropriately called the "Fruit-land Peninsula," and is very fertile, containing also celebrated sulphur and medicinal springs, which are much frequented by the sick from all the adjacent region. The county contains a great variety of soil, high and low hammock, high rolling pine land, much of which is heavily timbered with hickory, oak, and yellow pine. Many of the finest orange groves in the State are found here, and more than 5,000 acres are devoted to the cultivation of the orange alone, while cotton, rice, sugar, corn, sweet potatoes, and all the semi-tropical fruits form a permanent reliance for agricultural industry. Palatka, the county seat, is a very flourishing city, connected by rail and telegraph with all parts of the State which have railroad communications. San Mateo and Crescent City are flourishing towns, and the county is rapidly filling up with a good class of people from the Northern and Western States. The yearly farm productions amount to nearly $200,000, comprising principally cattle, butter, milk, swine, poultry, rice, grain, cotton, sugar, sweet potatoes, and honey.

GENERAL OBSERVATIONS.

This completes the list of localities visited by me, and they are in many respects the best counties in the State.

Except near the sea-coast marshes, mosquitoes are not very troublesome, and the noxious insects and reptiles avoid the settlements and generally confine themselves to the overflowed lands and swamps.

While much of the soil in this State is practically worthless in its present state, there are large tracts in the aggregate which are rendered very productive by the peculiar character of the climate and by the subterranean waters to which the roots of the trees find access, thereby causing the trees of the Citrus family to flourish even during protracted droughts which in the North would destroy every living thing.

There is a great demand for a government experiment station in Florida, which would be of incalculable service in showing the people what can be raised here and the best methods of culture.

The tea plants sent to this State by the Department of Agriculture have not received proper care; but the few that have survived the neglect demonstrate the fact that tea can be grown here successfully by intelligent culture.

It is believed by thoughtful men that many tropical products can be raised here in sufficient quantity to save to the nation many millions of dollars annually which are now sent to foreign countries. I refer especially to tea, coffee, sugar. silk, and the many articles of commerce now imported from the Indies, China, and Japan.

Colored laborers are hired here at wages averaging about $1 per day, without board.

Orange trees four years from the seed and one year from the bud will bear fruit in from three to four years.

Orange trees can be bought at prices ranging from twenty-five to seventy-five cents each, and they should be set out during the months of November, December, or January.

Lemon trees bear fruit in four years from the setting.

Bananas are produced in eighteen months from the setting, and ripen from June to December.

The transportation of a box of oranges from Sanford to New York costs about 45 cents.

It is a curious fact that while oranges begin to ripen in October, they may be left upon the tree without material deterioration for twelve months after they are ready for use.

Hoping this report may be of service to the most important of all our government departments, as well as to the interests of agriculture at large,

I remain yours, very truly,

JAMES H. FOSS.